T0344455

Advanced Materials towards Energy Sustainability

Industry 4.0 is revolutionizing the way companies manufacture, improve, and distribute their products. It demands the application of renewable energy using advanced materials. Renewable energy is reshaping the fields of industry, agriculture, and households, providing reliable power supplies and fuel diversification. This enhances energy security, lowers the risk of fuel spills, and reduces the need for imported fuels. Examples of material applications used for renewable energy are photovoltaic, solar cells, which can be used in agriculture. This volume has a diverse audience, including students, researchers, and academics engaged in materials and renewable energy.

Features

- Presents latest research on renewable energy in relation to urbanization, industrialization, and the environment.
- Provides in-depth discussion on modeling and simulation using latest techniques.
- Provides technical exposure for the readers on advanced materials.
- Provides numerous examples on properties of biomaterials and their future prospect.
- Provides up-to-date information on functional materials for industrial application.

Urbanization, Industrialization, and the Environment

Series Editor Viduranga Waisundara

Advanced Materials towards Energy Sustainability

Theory and Implementations

Edited By
Samsul Ariffin Abdul Karim

*Faculty of Computing and Informatics and Data Technologies
and Applications (DaTA) Research Lab,
Universiti Malaysia Sabah, Malaysia
Creative Advanced Machine Intelligence (CAMI) Research Centre,
Universiti Malaysia Sabah, Malaysia*

Poppy Puspitasari

*Department of Mechanical and Industrial Engineering
Universitas Negeri Malang
Centre of Advanced Materials for Renewable Energy (CAMRY)*

CRC Press
Taylor & Francis Group
Boca Raton London New York

CRC Press is an imprint of the
Taylor & Francis Group, an **informa** business

First edition published 2024
by CRC Press
6000 Broken Sound Parkway NW, Suite 300, Boca Raton, FL 33487–2742

and by CRC Press
4 Park Square, Milton Park, Abingdon, Oxon, OX14 4RN

Library of Congress Cataloging-in-Publication Data
Names: Abdul Karim, Samsul Ariffin, editor. | Puspitasari, Poppy, editor.
Title: Advanced materials towards energy sustainability : theory and implementations / Samsul Ariffin Abdul Karim, Faculty of Computing and Informatics and Data Technologies and Applications (DaTA) Research Group, Universiti Malaysia Sabah, Malaysia, Poppy Puspitasari, Department of Mechanical Engineering, Malang State University, Indonesia.
Description: First edition. | Boca Raton : CRC Press, [2023] | Includes bibliographical references and index.
Identifiers: LCCN 2023004173 (print) | LCCN 2023004174 (ebook)
Subjects: LCSH: Materials—Technological innovations. | Green products. | Renewable energy sources—Materials.
Classification: LCC TA403.8 .A38 2023 (print) | LCC TA403.8 (ebook) | DDC 338.2/60286—dc23/eng/20230222
LC record available at https://lccn.loc.gov/2023004173
LC ebook record available at https://lccn.loc.gov/2023004174

ISBN: 978-1-032-43499-5 (HB)
ISBN: 978-1-032-43538-1 (PB)
ISBN: 978-1-003-36781-9 (EB)

DOI: 10.1201/9781003367819

Typeset in Times
by Apex CoVantage, LLC

Contents

Preface

Energy is one of the most pressing issues confronting practically every country on the planet. This is because energy is one of the most important aspects in a country's economic development. When the increasing demand for energy from all countries across the world to fund their economic growth actually causes the supply of conventional energy reserves to diminish, energy challenges become more complicated. The total energy consumption for the globe currently stands at 10 terrawatts (equal to $3 \times 1,020$ joules/year), and it is expected to continue to rise until it reaches 30 terrawatts by 2030 [1–3]. In fact, the growing demand for energy collides with people's desire to live in a clean, pollution-free environment. These different issues necessitate the development of alternative energy sources capable of meeting the aforementioned problems. A solar cell is a type of power plant that converts sunlight into electricity. Solar energy is the most promising energy source due to its long-term sustainability and abundance. After many conventional energy sources are reduced in number and are not environmentally friendly, the sun is an energy source that is believed to be able to solve the problem of future energy needs. A total energy demand of 10 TW is $3 \times 1,020$ J per year.

The total solar energy reaching Earth's surface each year is $2.6 \times 1,024$ joules. In comparison, the total amount of energy that can be converted through photosynthesis on the entire surface of the globe each year is $2.8 \times 1,021$ J. When we compare the amount of energy required to the amount of solar energy arriving at the Earth's surface, we can meet all the world's energy needs by covering 0.05% of Earth's surface area (the total surface area of Earth is 5.1×108 km^2) with a solar cell with a 20% efficiency. Solar cells are a very good potential for an alternative energy source in the future because of the vast amount of energy provided by sunshine. Solar cells also have the advantage of being a practical energy source because they do not require any maintenance, because they can be placed in a modular fashion anywhere they are needed. Solar cells, unlike wind power plants, do not produce excessive noise and can be erected in almost any location because nearly every location in this portion of the world receives sunshine. In contrast, a water generator (hydro) can only function in places with a specified volume of water flow. With these advantages, it's no surprise that industrialized countries are competing to develop solar cells and solar cell manufacturing technology that are economically beneficial. The entire electrical energy generated by solar cells around the world has only reached roughly 12 GW (compared to the world's total power use of 10 TW) until today. Out of the 12 GW, Germany has installed the largest solar cell, which is almost 5 GW. However, every year, there is an increase in the production of solar cells, where, in 2008, the total production of solar cells worldwide reached 6.22 GW. This increasing production value is also followed by efforts to reduce the price of solar modules per watt peak. Currently, the price of electricity produced by solar cells is 50 cents per kWh, which is still relatively high when compared to generation from other sources, such as from thermal generators, which is only 8 cents per kWh.

Various technologies have been developed in the process of making solar cells to reduce production prices and be more economical. The types of solar cells that have now developed are based not only on crystalline silicon semiconductors but various types, ranging from thin layer, organic, single layer, and multi junction, to the newest type of dye-sensitized solar cell.

The development of advanced materials in the field of renewable energy is one of the focuses of development in various countries. This is because the energy problem is a global and national problem that must be solved systematically. The use of renewable energy is one of the more environmentally friendly solutions to replace energy sources from fossil fuels. Industry 4.0 demands the use of renewable energy based on the use of advanced materials. Renewable energy has touched all lines of life in the fields of industry, agriculture, and households. Renewable energy provides reliable power supplies and fuel diversification, which enhance energy security, lower the risk of fuel spills, and reduce the need for imported fuels. Renewable energy also helps conserve the nation's natural resources. The use of advanced materials as materials for renewable energy sources has positive values and impacts economically, environmentally, and also industrially. Some of the material applications used for renewable energy are materials for photovoltaic, solar cells, and even for agriculture. Based on this, seven chapters in this book have been written by authors competent in their fields. This book also relates to the Sustainable Development Goals (SDGs) of the United Nations, especially SDG 7, about affordable and clean energy, and SDG 9, about industry, innovation, and infrastructure. According to SDG 7, the lack of access to energy supplies and transformation systems is a constraint to human and economic development. The environment provides a series of renewable and non-renewable energy sources. Energy efficiency and increased use of renewables contribute to climate change mitigation and disaster risk reduction. Meanwhile, SDG 9 describes that the number of people employed in renewable energy sectors is presently around 2.3 million. Given the present gaps in information, this is no doubt a very conservative figure. Because of the strong rising interest in energy alternatives, the possible total employment for renewables by 2030 is 20 million jobs. The increasing need for renewable energy requires an increase in the use of advanced materials. Therefore, the research on functional materials to support alternative energy is very important. Several chapters in this book discuss functional materials developed to increase the effectiveness and efficiency of renewable energy.

We would like to thank all contributors, who provided an excellent contribution to this book. We are forever grateful for their commitment and contribution to this book. Special thank-you to the staff at Taylor & Francis/CRC Press for the publication of the book. The first editor is fully supported by the Faculty of Computing and Informatics, Universiti Malaysia Sabah.

The book is suitable for all postgraduates and researchers working in this rapidly growing research area.

Samsul Ariffin Abdul Karim,
Kota Kinabalu, Malaysia

Poppy Puspitasari,
Malang, Indonesia

Editor Biographies

Samsul Ariffin Abdul Karim is Associate Professor of the Software Engineering Programme, Faculty of Computing and Informatics, Universiti Malaysia Sabah (UMS), Malaysia. He is Core Member of Data Technologies and Applications (DaTA) Research Group, Faculty of Computing and Informatics, Universiti Malaysia Sabah. He obtained his PhD in mathematics from Universiti Sains Malaysia (USM). He is Professional Technologist registered with the Malaysia Board of Technologists (MBOT), No. Perakuan PT21030227. His research interest includes numerical analysis, machine learning, approximation theory, optimization, science, and engineering education, as well as wavelets. He has published more than 160 papers in journals and conferences, including three edited conference volumes, and 80 book chapters. He was the recipient of the Effective Education Delivery Award and Publication Award (journal and conference paper), UTP Quality Day 2010, 2011, and 2012, respectively. He was Certified Wolfram Technology Associate, Mathematica Student Level. He has also published 13 books with Springer Publishing, including six books with the *Studies in Systems, Decision and Control* (SSDC) series, two books with Taylor & Francis/CRC Press, one book with IntechOpen, and one book with UTP Press. Recently, he received the Book Publication Award in UTP Quality Day 2020 for the book *Water Quality Index (WQI) Prediction Using Multiple Linear Fuzzy Regression:Case Study in Perak River, Malaysia*, that was published by *SpringerBriefs in Water Science and Technology* in 2020.

Poppy Puspitasari is Associate Professor at Universitas Negeri Malang, Indonesia. She has worked as Lecturer in Mechanical Engineering Department since 2008. She finished her Ph.D at Universiti Teknologi Petronas, Malaysia. She has published more than 200 papers in journals and conferences, and 10 books about nanomaterials and their applications. She has a Scopus h-index of 11 and a Google Scholar h-index of 15, with 10 Indonesian patents. Her research interests are application of nanomaterials in Mechanical Engineeering fields such as catalyst for pyrolysis, nanofluid for heat exchanger, and nanolubricant. She has received several awards, such as gold medals in edX UTP, ITEX UTP, and iENA Germany. She also received awards as the best lecturer in Indonesia and at Universitas Negeri Malang.

Contributors

Hamidatu Alhassan
Applied Physics Department,
  Faculty of Science, Universiti
  Brunei Darussalam, Bandar Seri
  Begawan, Jalan Tungku Link,
  Gadong BE1410, Brunei

Nur Liyana Asnan
Mathematical Sciences Programme,
  Faculty of Science, Universiti
  Brunei Darussalam, Tungku Link
  Road, Gadong BE1410, Brunei

Mimi Asyiqin Asrahwi
Department of Chemistry, Faculty
  of Science, Universiti Brunei
  Darussalam, Jalan Tungku Link,
  Gadong BE1410, Brunei

Muhammad Roil Bilad
Faculty of Integrated Technologies,
  Universiti Brunei Darussalam,
  Jalan Tungku Link,
  Gadong BE1410, Brunei

Rosmaya Dewi
Faculty of Integrated Technologies,
  Universiti Brunei Darussalam,
  Jalan Tungku Link,
  Gadong BE1410, Brunei

Jonathan Hobley
Department of Biomedical
  Engineering, National Cheng Kung
  University, No. 1 University Road,
  Tainan City 701, Taiwan

Saiful Azmi Husain
Mathematical Sciences Programme,
  Faculty of Science, Universiti
  Brunei Darussalam, Tungku Link
  Road, Gadong BE1410, Brunei

Sridhar Sripadmanabhan Indira
Clean Technology Lab, Taylor's
  Lakeside Campus, No. 1 Jalan
  Taylor's, 47500 Subang Jaya,
  Selangor, Malaysia

**Nurulizzatul Ningsheh Mohammad
Shahri**
Department of Chemistry, Faculty
  of Science, Universiti Brunei
  Darussalam, Jalan Tungku Link,
  Gadong BE1410, Brunei

Taqiyyuddin Akram Aidil
Department of Chemistry, Faculty
  of Science, Universiti Brunei
  Darussalam, Jalan Tungku Link,
  Gadong BE1410, Brunei

Junaidi H. Samat
Department of Chemistry, Faculty
  of Science, Universiti Brunei
  Darussalam, Jalan Tungku Link,
  Gadong BE1410, Brunei

Cristina Pei Ying Kong
Department of Chemistry, Faculty
  of Science, Universiti Brunei
  Darussalam, Jalan Tungku Link,
  Gadong BE1410, Brunei

Nurul Amanina A. Suhaimi
Department of Chemistry, Faculty
of Science, Universiti Brunei
Darussalam, Jalan Tungku Link,
Gadong BE1410, Brunei

Samsul Ariffin Abdul Karim
Software Engineering Programme,
and Data Technologies and
Applications (DaTA) Research
Lab, Faculty of Computing and
Informatics, Universiti Malaysia
Sabah, Jalan UMS88400 Kota
Kinabalu, Malaysia
Creative Advanced Machine
Intelligence (CAMI) Research
Centre Universiti Malaysia Sabah,
Jalan UMS88400 Kota Kinabalu,
Malaysia

Chong Kok Keong
Department of Electrical and
Electronic Engineering, Lee Kong
Chian Faculty of Engineering
and Science, Universiti Tunku
Abdul Rahman, Jalan Sungai
Long, Bandar Sungai Long, 43000
Kajang, Selangor

Eny Kusrini
Department of Chemical Engineering,
Faculty of Engineering, Universitas
Indonesia, Kampus Baru UI,
Depok 16424, Indonesia

Lee Hoon Lim
Department of Chemistry, Faculty
of Science, Universiti Brunei
Darussalam, Jalan Tungku Link,
Gadong BE1410, Brunei

Abdul Hanif Mahadi
Centre for Advanced Material and
Energy Sciences, Universiti Brunei
Darussalam, Jalan Tungku Link,
Gadong BE1410, Brunei

Aisyah Farhanah Abdul Majid
Department of Chemistry, Faculty
of Science, Universiti Brunei
Darussalam, Jalan Tungku Link,
Gadong BE1410, Brunei

Nur Alimatul Hakimah Narudin
Department of Chemistry, Faculty
of Science, Universiti Brunei
Darussalam, Jalan Tungku Link,
Gadong BE1410, Brunei

Muhammad Nur
Center for Plasma Research,
Integrated Laboratory, Universitas
Diponegoro, Tembalang Campus,
Semarang 50275, Indonesia

Riana Nurmalasari
Department of Mechanical and
Industrial Engineering, Faculty of
Engineering, State University of
Malang, Jl. Semarang 5,
Malang 65145, Indonesia

Avita Ayu Permanasari
Department of Mechanical and
Industrial Engineering, Faculty
of Engineering; and Center of
Advance Materials and Renewable
Energy, State University of
Malang, Jl. Semarang 5, Malang
65145, Indonesia

Diki Dwi Purnomo
Department of Mechanical and
Industrial Engineering, Faculty of
Engineering, State University of
Malang, Jl. Semarang 5,
Malang 65145, Indonesia

Poppy Puspitasari
Department of Mechanical and Industrial
Engineering, Faculty of Engineering;
and Center of Advance Materials and
Renewable Energy, State University

of Malang, Jl. Semarang 5,
Malang 65145, Indonesia

Chandra Setiawan Putra
Department of Mechanical and
    Industrial Engineering, Faculty of
    Engineering, State University of
    Malang, Jl. Semarang 5,
    Malang 65145, Indonesia

Nurul 'Aqilah Rosman
Department of Chemistry, Faculty
    of Science, Universiti Brunei
    Darussalam, Jalan Tungku Link,
    Gadong BE1410, Brunei

Mohd Syaadii Mohd Sahid
Department of Chemistry, Faculty
    of Science, Universiti Brunei
    Darussalam, Jalan Tungku Link,
    Gadong BE1410, Brunei

Ensan Waatriah E. S. Shahrin
Department of Chemistry, Faculty
    of Science, Universiti Brunei
    Darussalam, Jalan Tungku Link,
    Gadong BE1410, Brunei

Norazanita Shamsuddin
Faculty of Integrated Technologies,
    Universiti Brunei Darussalam,
    Jalan Tungku Link,
    Gadong BE1410, Brunei

Cai Jen Sia
Department of Chemistry, Faculty
    of Science, Universiti Brunei
    Darussalam, Jalan Tungku Link,
    Gadong BE1410, Brunei

Ramsundar Sivasubramaniyam
Clean Technology Lab, Taylor's
    Lakeside Campus, No. 1 Jalan
    Taylor's, 47500 Subang Jaya,
    Selangor, Malaysia

Fredolin Tangang
Department of Earth Sciences and
    Environment, Faculty of Science
    and Technology, Universiti
    Kebangsaan Malaysia (The
    National University of Malaysia),
    Bangi 43600, Selangor, Malaysia

Anwar Usman
Department of Chemistry, Faculty
    of Science, Universiti Brunei
    Darussalam, Bandar Seri
    Begawan, Jalan Tungku Link,
    Gadong BE1410, Brunei

**Chockalingam Aravind
Vaithilingam**
Clean Technology Lab, Faculty
    of Innovation and Technology,
    Taylor's University, 47500 Subang
    Jaya, Selangor, Malaysia

Yvonne Soon Ying Woan
Applied Physics Department, Faculty
    of Science, Universiti Brunei
    Darussalam, Bander Seri Begawan,
    Jalan Tungku Link,
    Gadong BE1410, Brunei

Voo Nyuk Yoong
Applied Physics Department, Faculty
    of Science, Universiti Brunei
    Darussalam, Bander Seri Begawan,
    Jalan Tungku Link,
    Gadong BE1410, Brunei

Ten Ru Yuh
Mathematical Sciences Programme,
    Faculty of Science, Universiti
    Brunei Darussalam, Tungku Link
    Road, Gadong BE1410, Brunei

1 Introduction

Samsul Ariffin Abdul Karim
and Poppy Puspitasari

CONTENTS

1.1 INTRODUCTION

Advanced materials, such as metal-organic frameworks (MOFs), covalent-organic frameworks (COFs), MXene, and graphene oxide (GO), have the potential to greatly improve the efficiency, cost, and cleanliness of energy technologies. These materials can be used in a wide range of applications, including energy storage, conversion, generation, harvesting, and transport, as well as environmental remediation.

The use of advanced materials in energy and environmental applications requires a multi-disciplinary approach, bringing together experts from fields such as materials science, chemistry, biology, and physics. By combining their knowledge and expertise, researchers can develop new materials that are able to tackle key global energy and environmental challenges.

One potential application of advanced materials is for renewable and sustainable solar energy, such as photovoltaic (PV) and water treatment technology. These materials can also be used in the development of clean technologies for energy storage, agriculture, and the environment. The ultimate goal is to use advanced materials to produce energy in an environmentally friendly manner while also protecting our environment for future generations.

The scope of research in this area is broad and involves researchers from a variety of disciplines. This book is dedicated to showcasing new and original research on the use of advanced materials in energy and environmental applications, from both experimental and theoretical perspectives. By bringing together researchers from different fields, we can gain a better understanding of the potential of advanced materials and how they can be used to address the energy challenges and environmental issues. This book also relates to the United Nations Sustainable Development Goals (SDGs) [1], especially SDG 7, about affordable and clean energy, and SDG 9, about industry, innovation, and infrastructure. According to SDG 7, the lack of access to energy supplies and transformation systems is a constraint to human and economic development. The environment provides a series of renewable and non-renewable energy sources. Energy

DOI: 10.1201/9781003367819-1

efficiency and increased use of renewables contribute to climate change mitigation and disaster risk reduction. Meanwhile, SDG 9 describes that the number of people employed in renewable energy sectors is presently around 2.3 million. Given the present gaps in information, this is no doubt a very conservative figure. Because of the strong rising interest in energy alternatives, the possible total employment for renewables by 2030 is 20 million jobs. The increasing need for renewable energy requires an increase in the use of advanced materials. Therefore, the research on functional materials to support alternative energy is very important. Several chapters in this book discuss functional materials developed to increase the effectiveness and efficiency of renewable energy.

1.2   SUMMARIES

The following paragraphs provide the main summaries for each contributing chapter. Each contributed chapter can be regarded as a self-standing contribution.

Chapter 2, titled "Physical and Magnetic Properties of Nickel Cobalt Oxide ($NiCo_2O_4$) as a Promising Material for Supercapacitor Synthesized by Self-Combustion Method," discusses the need for energy storage devices with large capacities and long technical life, which becomes important. A supercapacitor is one of the energy storage devices used in the telecommunications world and is increasingly entering the automotive world as a fuel saver. Some of the advantages of supercapacitors over batteries and capacitors are their high power, long technical life, and simple design. Currently, the manufacture of supercapacitors is still dominated by materials made from carbon, polymers, and transition metal compounds, such as graphene, CNT, biomass, PANI, MnO_2, and TiO_2. These materials tend to have a high price with a complicated synthesis process; for this reason, it is necessary to have nickel-based or cobalt-based raw materials with lower prices and an easy synthesis process. Nickel cobalt raw materials are also more abundant, with easier oxidation and reduction conditions. Nickel cobalt oxide ($NiCo_2O_4$), in this study, was synthesized using the self-combustion method, which was then characterized for its physical properties in the form of phase identification, morphology, elemental analysis, material functional groups, and also electrical properties as a dielectric material.

Chapter 3, titled "Variety of Agricultural Machine Innovations by Utilizing Renewable Energy," discusses the vital sector of the Indonesian economy as an agricultural country that contributes significantly to food security. As a result, it is reasonable for the agricultural sector to continue to develop and receive government attention. Numerous obstacles must be overcome and exploited to the fullest extent possible. To maximize agriculture's potential in Indonesia, innovation is required. The innovations that have the potential to be developed are related to agricultural machines that utilize renewable energy to increase agricultural productivity. In Indonesia, the use of renewable energy sources, such as solar cells, wind, and heat, presents significant opportunities.

Chapter 4, titled "Off-Axis Reflectors for Concentrated Solar Photovoltaic Applications," discusses the application of off-axis reflectors for concentrated photovoltaics applications. Off-axis reflectors have mostly only been deployed for optical and lighting applications. In this work, the potential of off-axis reflectors for solar photovoltaics is investigated via ray-tracing studies carried out via the Monte Carlo method. Off-axis reflectors prove to be a new approach to improving the applicability of concentrated photovoltaic technologies.

Chapter 5, titled "Modeling the Feasibility of Using Solar PV Technology for Residential Homes in Brunei," discusses solar PV modules to discover the most convenient and outstanding achievement of solar PV technologies for local meteorological conditions. There are six types of modules: (i) monocrystalline silicon (m-cSi), (ii) polycrystalline silicon (p-cSi), (iii) amorphous silicon (a-Si), (iv) multi-junction silicon (tandem), (v) copper-indium-selenium (CIS), and (vi) heterojunction with an intrinsic thin layer (HIT). All module types have been installed with a nominal capacity of 1.2 kWp, in which the solar farm covers an area of about 12,000 square meters, with exactly 9,234 pieces of solar panels. This chapter will discuss the feasibility of using solar PV technology for residential homes in Brunei using Photovoltaic Geographical Information System (PVGIS) modeling by estimating its energy output and irradiation received by the previously mentioned types of solar panels for selected on-site locations in Brunei, with the given optimal solar PV performance criteria as well as rooftop solar PV design considerations. Limitations and challenges of this study will also be discussed in this chapter.

Chapter 6, titled "Synthesis Techniques and Applications of Graphene and Its Derivatives," discusses the synthesis and functionalization techniques of graphene and graphene-derived nanomaterials that utilized both organic and inorganic carbon sources. The potential industrial application of the synthesis processes is discussed and evaluated on variables such as product quality, scalability, cost, and environmental impact.

Chapter 7, titled "Chitin and Chitosan: Isolation, Deacetylation, and Prospective Biomedical, Cosmetic, and Food Applications," discusses chitin that is mainly isolated from shrimp, crab, and other crustacean shells. Chitin is deacetylated using strong basic solutions into chitosan, and the kinetics of heterogeneous deacetylation process follows a first-order kinetics. This study summarizes isolation of α-chitin from the crustacean shells. The chemical and crystalline structures of the α-chitin before and after deacetylation process are characterized by FTIR spectroscopy and XRD analysis. The effects of temperature and basic concentration on the deacetylation kinetics are evaluated and discussed in detail.

Chapter 8, titled "Quantum Dots: From Synthesis to Biomedical and Biological Applications," discusses the importance of capping agents, which are necessary to stabilize quantum dots. In fact, the capping agents not only play an important role in passivating metallic atoms on QD surfaces but also provide an extensive way to modify the QD surface to be conductive, facilitating charge transport and enhancing interdot coupling. With their distinctive photostability, distinctive design and engineering of capping agents, and rich surface chemistry, QDs can be used as traceable drug nanocarriers in drug delivery systems. Specific structural features of capping agents can also be used to alter the biological activities of QDs for visualization, diagnostic, labeling, and therapeutic purposes. This chapter summarizes several capping agents used in the fabrication of nanoparticles and QDs, as well as the recent advancements of the resulting QDs applicable for biological and biomedicinal applications. Finally, future directions of the capping agents for biological and biomedicinal applications are also discussed.

Chapter 9, titled "Biopolymers: Sources, Chemical Structures, and Applications," discusses the chemical structures, important characteristic, and applications of biopolymers, especially those belonging to polyamides, polysaccharides, polyesters, polycarbonates, and vinyl polymers. The rich amide, amino, acetyl, hydroxyl, and sulfonate functional groups interconnect the biopolymer strands through hydrogen bonding interactions and stabilize their three-dimensional structures. In general, the biopolymers are biocompatible and biodegradable, so they are fascinating advanced materials for food, food packaging, agriculture, wastewater treatment, cosmetic, biomedical, and high-technology applications.

Finally, in Chapter 10, we elaborate further on the "Conclusion and Future Prospects" of advanced materials.

1.3   ACKNOWLEDGMENTS

Special thank-you to the Faculty of Computing and Informatics, Universiti Malaysia Sabah, for the financial and computing facilities support that has made the completion of the book possible.

REFERENCE

[1] https://sdgs.un.org/goals (Retrieved on 22 December 2022).

2 Physical and Magnetic Properties of Nickel Cobalt Oxide (NiCo$_2$O$_4$) as Synthesized by Self-Combustion Method

*Poppy Puspitasari, Chandra Setiawan Putra,
Avita Ayu Permanasari, Diki Dwi Purnomo,
and Samsul Ariffin Abdul Karim*

CONTENTS

DOI: 10.1201/9781003367819-2

2.1   INTRODUCTION

According to [1], energy demand in 2050 will reach 2.9 billion barrels of oil equivalent (BOE) in 2050. This figure increases from the 2040 projection of 2.1 billion BOE. The projected increase in energy demand is in accordance with economic growth, population, energy prices, and government policies. By sector, energy demand will be dominated by the industrial sector, with an estimated average growth of 3.9% per year. Then the commercial, household, and other sectors also continued to increase in line with the increase in the economy and population. Meanwhile, the growth rate of the transportation sector is estimated to be lower than the industrial sector; namely, 3.2% per year. By type, the final energy demand is still dominated by fuel oil (BBM), with an average growth rate of 2.8% per year. This happens because the use of fuel equipment technology is still more efficient than other energy equipment. In addition, biodiesel has also experienced an increase due to its role as a substitute for fuel, which can be used in several sectors, such as transportation, industry, commerce, and power generation. Furthermore, the development of electricity-based technological innovations in almost every sector, such as electric vehicles, will result in electricity demand increasing in 2050, with a growth rate of 4.7% per year.

Until now, Indonesia is still very dependent on fossil energy to meet the electricity needs of its population. Until the end of 2019, fossil energy still dominated the primary energy mix for power generation by 87.6%, leaving only 12.4% of the mix for renewable energy [2]. With the increasing demand for energy and growing concerns about air pollution and global warming, it is urgent to find alternative sources of energy to replace fossil fuels.

In today's modern life, electrical energy is a major need that cannot be avoided. Various technologies exist today, most of which require electrical energy storage devices. So far, there has been great interest among researchers in developing and improving more efficient energy storage devices. One such device is a supercapacitor.

Supercapacitors, also known as ultracapacitors or electrochemical capacitors, have attracted great attention due to their fast recharging capability, high power performance, long cycle life, and low maintenance costs. Supercapacitors utilize surface electrodes and thin dielectric electrolyte solutions to achieve capacitance several times greater than conventional capacitors [3]. Several metal oxide nanoparticle materials, such as RuO_2, NiO, MnO_2, Co_3O_4, and $NiCo_2O_4$ spinel, are considered very promising materials for supercapacitors because of their high specific capacitance, excellent cycle stability, low cost, and environmental friendliness. So metal oxide nanoparticles will have physical and chemical properties that are superior to the same material in large sizes, such as mechanical, electrical, magnetic, and optical strength [4].

NiCo$_2$O$_4$ nanoparticles are generally considered as mixed valence oxides that adopt a pure spinel structure in which nickel occupies octahedral sites and cobalt is distributed in both octahedral and tetrahedral sites, which can be applied in various fields, such as electrochromic films, fuel cell catalysts, gas sensors, cathode electrodes of batteries, magnetic materials, photovoltaic devices, electrochemical supercapacitors, and color-sensitive photocathodes. Research on various potential applications of NiCo$_2$O$_4$ magnetic nanoparticles continues to this day. One example is because of its good electron conductivity, stable structure, environmental friendliness, low cost, easy preparation, and much higher electrical conductivity than NiO and CoO. With the various advantages of the NiCo$_2$O$_4$ nanoparticle material, it is expected to be able to make supercapacitors more effective and efficient [5].

The superior properties and the wide application of NiCo$_2$O$_4$ nanoparticles in various aspects of life encourage the continued development of various synthesis methods in order to produce NiCo$_2$O$_4$ nanoparticles with the desired properties and characteristics. Particle size greatly influences the properties and characteristics of NiCo$_2$O$_4$ nanoparticles [6]. Thus, research studies on controlling the size of NiCo$_2$O$_4$ nanoparticle materials are urgently needed to determine their properties, characteristics, and potential applications. Controlling the size and properties of NiCo$_2$O$_4$ nanoparticles can be carried out during the material synthesis process.

Synthesis of nickel cobalt oxide nanoparticles was carried out using the self-combustion method to produce nano-sized particles and to see the characteristics of the material properties. Nanoparticles synthesized by this method are predicted to produce many products with high purity and smaller crystal size [7]. Furthermore, this research studied the variation of solvents and fuels used in the self-combustion synthesis method on particle size, crystal structure, and magnetic properties of nickel cobalt oxide (NiCo$_2$O$_4$) nanoparticles. Physical phenomena related to the crystal structure, grain size, and magnetic properties of nanoparticles are discussed based on data from the characterization of XRD (X-ray diffraction), SEM (scanning electron microscopy), FTIR (Fourier-transform infrared spectroscopy), and VSM (vibrating-sample magnetometer).

2.2 LITERATURE REVIEW

2.2.1 NICKEL OXIDE (NiO)

The very unique properties of nanomaterials have attracted technological and industrial interest; one example of which is nickel cobalt oxide (NiCo$_2$O$_4$) nanomaterials. NiCo$_2$O$_4$ metal oxide has properties—namely, optical, electronic, thermal, and magnetic properties—which are related to its general characteristics, such as mechanical hardness, thermal stability, or chemical passivity [8].

Based on these unique properties, $NiCo_2O_4$ can be applied in various fields, such as electrochromic films, fuel cell catalysts, gas sensors, battery cathodes, magnetic materials, photovoltaic devices, electrochemical supercapacitors, and color-sensitive photocathodes. Most $NiCo_2O_4$ applications require better material properties and characteristics [9].

NiO nanoparticles are transition metal oxides that have an FCC (face-centered cubic) crystal structure with a density of 6.67 g/cm^3, a melting point of 1,955°C, and self-ignition at a temperature of 400°C (Figure 2.1). NiO is a material with wide technological potential and has been studied as a p-type semiconductor with a wide and stable bandgap value of 3.6–4.0 ev [10]. NiO has a size range of 4 nm–40 nm and has certain magnetic properties, such as superparamagnetism, superanti-ferromagnetism, or ferromagnetism, depending on the structure and synthesis method used. Meanwhile, large NiO particles are insulators with anti-ferromagnetic properties with a Neel (TN) temperature of 524 K [11].

Material synthesis methods need to be carried out to obtain NiO nanoparticles. Material synthesis is used to control the morphology, structure and size of the material in order to obtain NiO material with the desired properties and characteristics. The NiO synthesis process can be carried out using various methods, such as precipitation, thermal decomposition, heat treatment, solvothermal, and sol-gel methods [13].

2.2.2 COBALT OXIDE

Cobalt(II) oxide is an inorganic compound that has a visual appearance of olive green to red crystals, or grayish to black. Cobalt(II) oxide is usually applied as a pigment for coloring ceramics and paints; for drying paints, varnishes, and oils; for tinting glass, as a catalyst; and for preparing other cobalt salts. Cobalt oxide (Co_3O_4) has a normal spinel crystal structure based on the cubic arrangement of oxide ions, wherein Co(II) ions occupy the 8a tetrahedral state and Co(III) ions occupy the 16d octahedral site (Figure 2.2). Co_3O_4 is a p-type semiconductor with special properties due to its potential for use in sensors, heterogeneous catalysts, electrochemical devices, Li-ion batteries, and magnetic materials [14].

The commercial product is a cobalt oxide mixture, the commercial product is usually a dark gray powder, but the color may vary from olive oil to brown, depending on the particle size. Its density is 6.44 g/cm^3, which can also vary from 5.7 g/cm^3 to 6.7 g/cm^3, depending on the preparation method, and the melting point is around 1.830°C. It is insoluble in water but soluble in acids and alkaline [4]. Cobalt(II) oxide can absorb oxygen at normal temperatures. Heating at low temperatures with oxygen produces cobalt(III) oxide. Cobalt(II) oxide reacts with acids to form cobalt(II) salt. Reaction with sulfuric, 8hydrochloric, and nitric acids to yield sulfate, chloride, and nitrate, respectively, was obtained after evaporation of the solution [15].

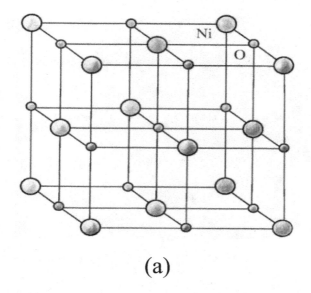

(a)

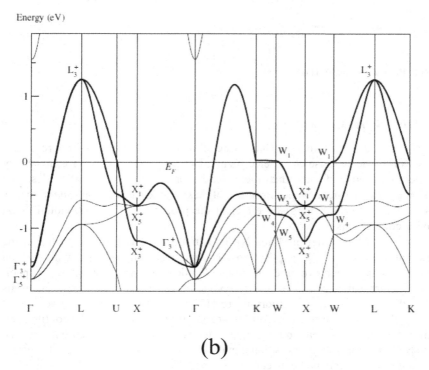

(b)

FIGURE 2.1 (a) Face-centered cubic on NiO and (b) Paramagnetic structure on conventional-band face-centered cubic of NiO [12].

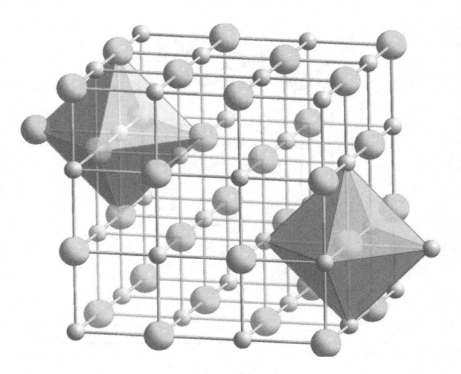

FIGURE 2.2　Cobalt(II) oxide molecular structure [14].

2.2.3　Nickel Cobalt Oxide

Nickel cobalt oxide ($NiCo_2O_4$) has attracted widespread attention from researchers worldwide because of its electrochemical properties, which are invariably superior to those of individual metal oxides or mixtures of metal oxides. The excellent electrochemical performance of these single-phase binary metal oxides is attributed to the synergistic effect of the properties of the individual metal oxide components [6].

Conjugated polymer-modified $NiCo_2O_4$ materials have been reported in the literature with versatile applications. N- and p-doped $NiCo_2O_4$ with oxygen vacancies have been explored for electrochemical performance for supercapacitors, electro-catalysts for O_2 and H_2 evolution reactions, and anodic materials for lithium-ion batteries. According to [16], they explored the arrangement of s-doped $NiCo_2O_4$ nanosheets as efficient and bifunctional electrodes for the overall water separation reaction. Compared to the $NiCo_2O_4$ not doped with metals, the $NiCo_2O_4$ doped with transition metals and earth metals is considered superior because of its excellent electrical conductivity.

The composite/microstructure of $NiCo_2O_4$ nanomaterials is also found to be suitable for potential applications in supercapacitors, fuel cells, Li-ion batteries,

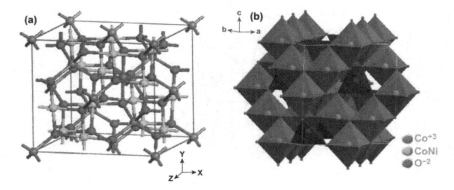

FIGURE 2.3   Molecular structure of NiCo$_2$O$_4$ [6].

electrocatalysts for oxygen reduction reactions and oxygen evolution reactions, photo-detectors, optoelectronic devices, perovskite solar cells, gas sensors, and biosensors (Figure 2.3) [6].

2.3 MAGNETIC PROPERTIES

Magnetic properties are properties possessed by a material that give the material a magnetic field so that it can attract objects made of metal that are nearby [9]. Based on its magnetic properties, the material can be divided into two types; namely, magnetic materials and non-magnetic materials. Magnetic materials contain metal elements, while non-magnetic materials are materials that do not have magnetic properties; for example, water, wood, and others.

Based on the origin of magnetic materials, they can be divided into two parts; namely, naturally and through manufacturing processes carried out by humans. Making magnetic materials can be done in three ways; namely, rubbing metal materials with permanent magnets, electromagnetism, and induction by permanent magnets (Figure 2.4) [17].

The magnetic properties are due to the direction of the magnetic moments on the individual electrons. Each electron in an atom has a magnetic moment that comes from two sources; namely, the movement of the atom around the atomic nucleus and the rotation of electrons around its axis. The movement of electrons around the atomic nucleus and the rotation of electrons about their axes will result in a spin magnetic moment. A material can have strong magnetic properties if the magnetic moments are arranged regularly (unidirectional). In non-magnetic materials, the magnetic moment has an arbitrary (irregular) direction, causing mutually exclusive effects [18].

The irregular magnetic moment causes non-magnetic materials to have no magnetic poles, while magnetic materials have two magnetic poles; namely, a

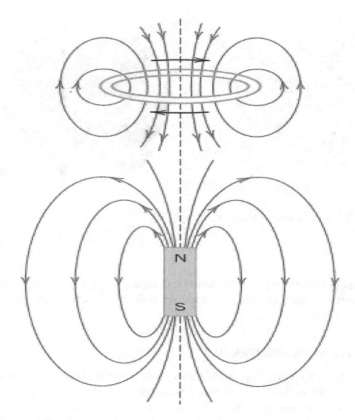

FIGURE 2.4 Magnetic force loop on magnetic materials [17].

north pole and a south pole. The magnetic poles are the areas at the ends of the magnets that have the greatest magnetic strength. The strength (intensity) of the magnetic field that each magnetic material has is different. According to international units (SI), the unit for magnetic intensity is the tesla (T), and the unit for total magnetic flux is the weber (Wb) (1 Wb/m^2 = 1 T), which is affected by the area in square meters [18]. According to Buck's (2017) book on the magnetic properties of the external magnetic field, materials can be divided into five types:

2.3.1 Diamagnetic

A diamagnetic material is a material in which the resultant atomic magnetic field of each atom or molecule is zero but the orbits and spins are not zero. Diamagnetic materials have negative susceptibility values (magnetic susceptibility/sensitivity). This material has unbalanced electron orbitals so that when it is given an external magnetic field, it will produce a magnetic moment in the

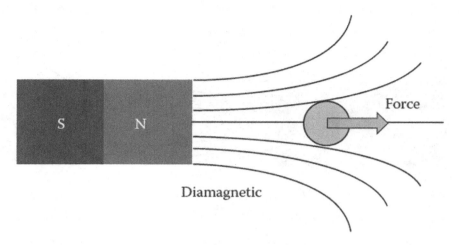

FIGURE 2.5 Diamagnetic materials without a magnetic moment [19].

opposite direction to that of the external magnetic field, which will produce a magnetic moment in the opposite direction to the external magnetic field, causing mutually canceling effects [19].

The diamagnetic properties of materials are caused by the rearrangement of electron orbitals under the influence of an external magnetic field. So it can be concluded that all materials have diamagnetic properties when influenced by an external field because all material atoms have electron orbitals. Diamagnetic material atoms do not have a magnetic moment, as shown in Figure 2.5.

A material can have a strong magnetic intensity if the arrangement of atoms in the material has unpaired electron spins. In a diamagnetic material, almost all electron spins are paired; as a result, this material does not attract lines of force, so the intensity of the magnetic field is very weak. Materials belonging to the diamagnetic group are generally non-magnetic materials, such as wood, water, organic compounds (petroleum), several types of polymers, and several metals (copper, mercury, gold, and bismuth).

2.3.2 PARAMAGNETIC

A paramagnetic material is a material that has a non-zero resultant atomic field value for each atom/molecule but the total atomic magnetic field resultant for all atoms in the material is zero. This is due to the magnetic moment on the paramagnetic material, which has an arbitrary direction, as shown in Figure 2.6.

Paramagnetic properties arise from the alignment of the magnetic moments of the electron spins that are in the same direction as the external magnetic field. If a paramagnetic material is affected by an external magnetic field, the electrons will change in such a way that the resultant atomic magnetic field is in the same

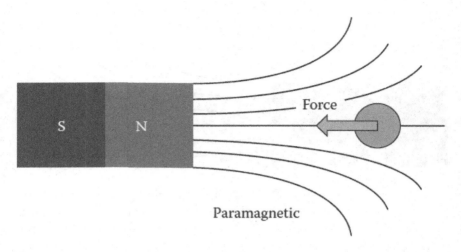

FIGURE 2.6 Direction of a magnetic moment on paramagnetic materials [19].

direction as the external field. This happens because of the torque resulting from the interaction between the electrons of the paramagnetic material and the external magnetic field.

Paramagnetic materials have a positive but very low susceptibility value. In a paramagnetic material, some atoms or ions have a net magnetic moment (zero value) because some unpaired electrons make this material slightly attract the lines of force so that it has magnetic intensity but cannot retain its magnetic properties when the external magnetic field is removed. Paramagnetic properties have weak attractive forces with magnets. Some examples of paramagnetic materials are aluminum (Al), magnesium (Mg), tungsten (W), molybdenum (Mo), lithium (Li), and tantalum (Ta).

2.3.3 FERROMAGNETIC

Ferromagnetic materials are materials that have a large atomic field resultant. This is mainly due to the magnetic moment the spin electrons have. The ferromagnetic properties arise due to the large number of electrons in the unpaired spins. If the ferromagnetic material is affected by an external magnetic field, then the electrons are unpaired. If the ferromagnetic material is affected by an external magnetic field. The more paired electrons, the greater the resulting magnetic intensity. In contrast to paramagnetic materials, these materials retain their magnetic properties when the external magnetic field is removed (the property of remanence). Therefore, this material is very good as a source of permanent magnets.

Ferromagnetic materials have a very large positive susceptibility value. The magnetic moments of ferromagnetic materials are arranged regularly

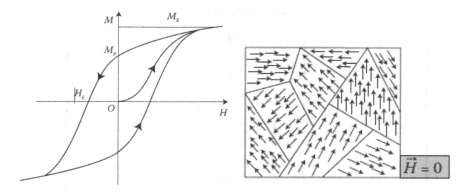

FIGURE 2.7   Direction of a magnetic moment on ferromagnetic materials [19].

(parallel), as shown in Figure 2.7. The magnetic moment in a ferromagnetic material is a net magnetic moment because it has many unpaired electrons. The magnetic intensity of the individual atoms in a ferromagnetic material is so strong that the interaction between the atoms will cause most of the atoms to align and form groups. Groups of atoms that align themselves in a region are called domains.

Materials belonging to the ferromagnetic group are generally magnetic materials, such as metals. This material has a strong magnetic intensity. However, when a ferromagnetic material is heated above a certain temperature, its ferromagnetic properties can change to paramagnetic properties. This temperature is called the Curie temperature. Each ferromagnetic material has a different Curie temperature; for example, a weak iron material has a Curie temperature of 770°C, and steel has a Curie temperature of 1,043°C.

2.3.4   ANTI-FERROMAGNETIC

Anti-ferromagnetic material is a magnetic material that has a low susceptibility value at all temperatures, with changes in temperature susceptibility occurring due to special circumstances. This material is only slightly affected by the presence of an external magnetic field [19]. Anti-ferromagnetic materials have atomic or molecular magnetic moments associated with electron spins that are arranged in a regular pattern with the other spins (on different sublattices) pointing in opposite directions (anti-parallel), as shown in Figure 2.8.

The regularity of magnetic moments in anti-ferromagnetic materials is generally at a fairly low temperature and disappears above a certain temperature. This temperature is known as Neel temperature. Neel temperature is the temperature that marks the change in magnetic properties from anti-ferromagnetic to paramagnetic. So if the anti-ferromagnetic material is heated above the Neel

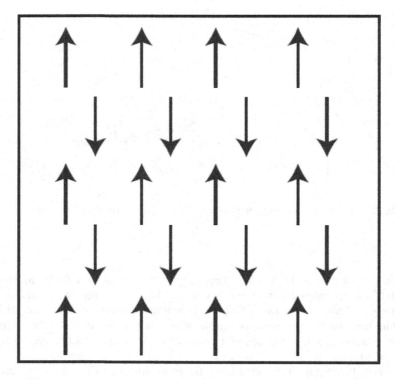

FIGURE 2.8   Direction of a magnetic moment on anti-ferromagnetic materials [19].

temperature, the anti-ferromagnetic material will become paramagnetic. This Neel temperature is the same as the Curie temperature in ferromagnetic materials. Anti-ferromagnetic materials have a low Curie temperature, which is only around 37°C.

An anti-ferromagnetic material can be described as a crystal structure with a lattice filled with two types of atoms with opposite (anti-parallel) magnetic moments. If there is no external magnetic field, the magnitude of the magnetic moments of the individual atoms in the anti-ferromagnetic material will be balanced so that the total magnetization is zero. Various oxide, sulfide, and chloride compounds are classified as antiferromagnetic, including nickel oxide (NiO), ferrous sulfide (FeS), cobalt chloride ($CoCl_2$), and manganese oxide (MnO) [20].

2.3.5 Ferrimagnetic

A ferrimagnetic material is a magnetic material that has a very large positive susceptibility value that depends on temperature. The magnetic domains in a ferrimagnetic material are divided into regions in opposite directions, but the

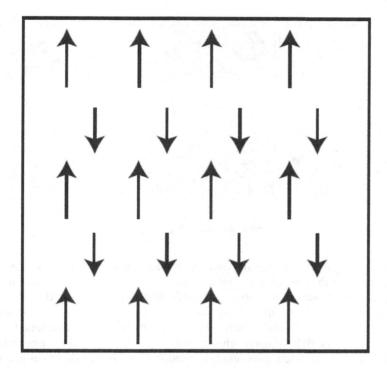

FIGURE 2.9 Ferrimagnetic magnetic moment direction [19].

total magnetic moment is not zero. Thus, it can be concluded that all magnetic minerals are ferrimagnetic materials. Although in some cases, the magnetization of rocks depends on the instantaneous strength of Earth's magnetic field around it and its magnetic mineral content.

Ferrimagnetic materials have a much higher resistivity value when compared to ferromagnetic materials. Resistivity is the ability of a material to conduct electric current, which depends on the magnitude of the electric field and current density. Ferrimagnetic properties are obtained when the ions in the material have magnetic moments that are aligned in an anti-parallel arrangement (opposite directions), as shown in Figure 2.9. Thus, the magnetic moment is not completely erased and the magnetization network will still exist, even though there is no influence from an external magnetic field (Halliday & Resnick, 2010).

2.4 NANOTECHNOLOGY

Nanotechnology is defined as the science and engineering that includes the design, synthesis, characterization, and application of materials that are organized in at least one dimension on a scale of nanometers or billions of meters.

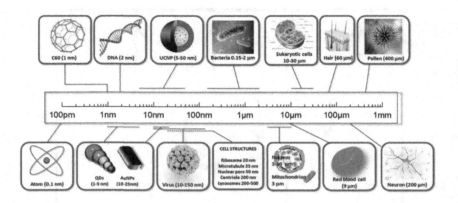

FIGURE 2.10 Application of nanomaterials [20].

Nanotechnology is the manipulation of materials that are ≤ 100 nm in size and at least fall into the one-dimensional category, where the physical, chemical, and biological properties are fundamentally different from bulk materials [21].

Nanotechnology has the ability to manipulate, control, and synthesize materials at the atomic and molecular levels to continually be developed. New discoveries in the field of nanotechnology emerge for new applications in various fields, such as electronics, energy, chemical, medical, and environmental (Figure 2.10). In the electronics field, nanotechnology is used for the development of nanometer-sized devices. In the energy sector, nanotechnology is used to manufacture more efficient solar cells. In the chemical field, nanotechnology is used for the development of more efficient catalysts and better-quality batteries [22]. In the medical field, nanotechnology is used to develop new tools for detecting cancer cells based on interactions between cancer cells and nanometer-sized particles, and developing drugs with nanometer-scale grains so they can dissolve and react more quickly in the body, as well as developing smart drugs (smart) that can search for tumor cells in the body and immediately kill these cells without disturbing normal cells. And in the environmental field, nanometer-sized materials are used to destroy organic pollutants in water and air [22].

The process of making the material into nanometers (nanoparticles) will cause two phenomena of change in the material; namely, the phenomenon of changes in the ratio of the number of atoms occupying the surface to the total number of atoms and the quantum phenomenon. The phenomenon of changes in the ratio of the number of atoms occupying the surface to the total number of atoms will affect changes in boiling point, freezing point, and chemical reactivity. While quantum phenomena will affect several material properties, such as electrical, optical, magnetic, and mechanical strength. Quantum phenomena occur as a result of limited space for electrons and other charge carriers in the particles. Changes in these properties can be controlled in the desired direction and can be an advantage of nanoparticles when compared to similar materials in bulk [22].

Nanoparticle materials have a number of chemical and physical properties that are superior to large materials. According to [23], some of the advantages of nanoparticle material properties, in general, include the following:

2.4.1 ELECTRICAL PROPERTIES

The effect of reducing the particle size on the electrical properties of nanoparticles can increase the conductivity of nanometals, generate nanodielectric conductivity, and increase the dielectric inductance for ferroelectrics.

Nanoparticles can have greater energy than ordinary-sized materials because they have a large surface area. The surface energy of the material will gradually increase due to changes in the electron orbitals. In general, the electrical resistivity of materials increases with decreasing material size.

2.4.2 OPTICAL PROPERTIES

The nanocrystalline nanoparticle system has interesting optical properties and is different from the crystalline system of large materials. The effect of reducing particle size on the optical properties of nanoparticles can increase the absorption (absorbance) of ultraviolet light, optical absorption oscillations, and the value of the bandgap.

2.4.3 MAGNETIC PROPERTIES

Magnetic strength is a measure of the degree of magnetism of a material. The effect of reducing particle size and increasing the ratio of surface area to the volume of particles on the magnetic properties of nanoparticles, among others, can increase or decrease the value of the magnetic coercivity of a material, reduce the Curie temperature of ferromagnetic materials, make nanoparticles have paramagnetic or ferromagnetic properties, make or increase the ability of nanoparticles, retain its magnetic properties (remanent magnetization), and increase magnetic permeability in ferromagnetic properties [22].

2.4.4 MECHANICAL PROPERTIES

The effect of reducing particle size on the mechanical properties of nanoparticles can increase the hardness, strength, fracture ductility, and wear resistance of the material. Nanoparticles have greater hardness and scratch resistance when compared to large-sized materials.

The properties of nanoparticles can be changed and regulated by controlling the size of the material, setting the chemical composition of the material constituents, modifying the surface, and controlling the interactions between material particles in the nanoparticle synthesis process. Nanoparticle synthesis means the process of making nanoparticle materials with a size of less than 100 nm and at the same time changing their properties or functions [20]. Nanoparticle synthesis can be carried out in various material phases, such as solid, liquid, or gas, using two methods:

2.4.4.1 Top-Down Method

The top-down method is a process of physically synthesizing nanoparticles in which there is a process of breaking large crystalline material particles into nanometer-sized material fragments (nanosizing). This technology uses various types of milling tools and homogenization techniques. An example of the top-down method is grinding with a milling tool, such as ball milling (Figure 2.11) [22].

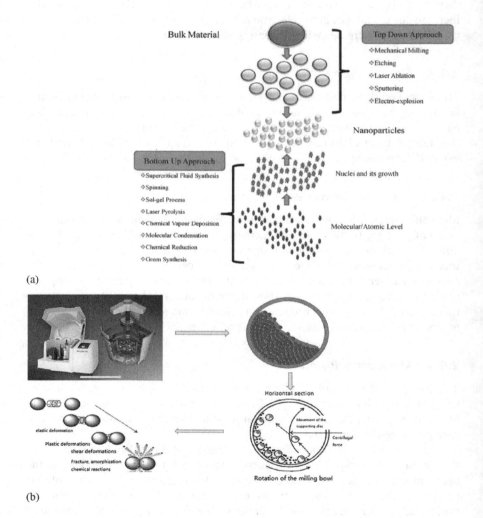

FIGURE 2.11 (a) Top-down and bottom-up method; (b) solid-state ball mill [20].

2.4.4.2 Bottom-Up Method

The bottom-up method is a process of chemically synthesizing nanoparticles by involving a chemical reaction from a number of precursors to produce new nanometer-sized materials. In the synthesis of nanoparticles, there are several factors that influence it; namely, the concentration of reactants, coating molecules (capping agent), temperature, and stirring. This method is a technique used to organize and control atoms and molecules to become nanometer-sized materials. Examples of bottom-up are the sol-gel synthesis method (Figure 2.12), coprecipitation, and gas phase agglomeration [20].

2.5 SELF-COMBUSTION

The self-combustion method is also known as self-propagating high-temperature synthesis (Figure 2.13). Self-combustion is one of the most versatile, convincing,

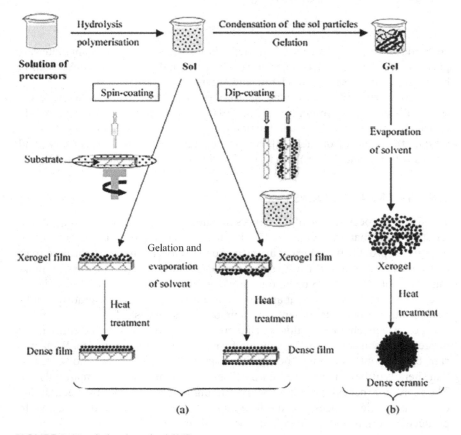

FIGURE 2.12 Sol-gel method [20].

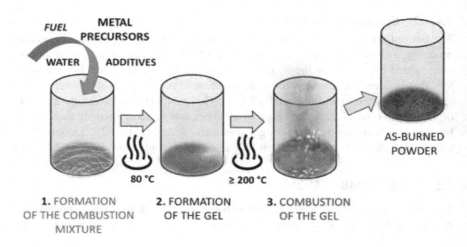

FIGURE 2.13   Self-combustion method.

convenient, cost-effective, and fast methods for the synthesis of nanomaterials [24]. The combustion mechanism involves a thermally induced redox reaction between precursor salts as oxidizers and organic fuels [25].

Glucose, fructose, tartaric acid, sucrose, glycine, citric acid, hydrazine, urea, and oxalic acid are commonly used as organic fuels. However, if metal oxalate or acetate salts are used, the combustion process can be carried out immediately without the presence of fuel [25]. By-product gases, such as CO_2, H_2O, N_2, and oxides N (NO_x) and S (SO_x), are formed during the combustion process [20].

2.6   STIRRING PROCESS

The stirring process is one of the factors that influence the properties and characteristics of the material synthesized using the sol-gel method [6]. The purpose of the stirring process is for the occurrence of perfect mixing between the precursor and solvent in a solution so as to speed up the chemical reaction process for making new materials and to be able to produce materials with small and uniform sizes. Mixing is a process that aims to reduce differences in temperature conditions or other properties present in precursors and solvents [6].

The basic mechanism of this stirring process involves moving the entire mass of precursor and solvent in the solution by means of a rotating device. With stirring by the tool, movement occurs in all precursor and solvent molecules so that there will be contact between particles to react to form new materials [3]. Therefore, it is necessary to control the stirring process in order to speed up the chemical reaction process for the formation of new materials and to be able to produce materials with smaller and more uniform sizes [3].

The variation of stirring time can affect the precursor and solvent mixing process, which will affect the level of speed and perfection of the chemical reaction for making the material and the size of the material produced. The provision of excessive stirring time can damage the resulting material product. Each material has a different optimum stirring time in order to produce the desired level of particle homogeneity [26].

The homogeneity of the solution causes the nucleation process to occur evenly throughout the material point, thus forming a small particle size and almost uniform. The duration of stirring causes the formed NiCo$_2$O$_4$ particles to rebond with oxygen or lose oxygen. If the NiCo$_2$O$_4$ particles bind to oxygen, the NiCo$_2$O$_4$ particles will be negatively charged and vice versa will be positively charged. The positively and negatively charged NiCo$_2$O$_4$ particles will attract each other and will eventually combine to form larger NiCo$_2$O$_4$ particles. This situation is not completely homogeneous, depending on the length of the stirring process [13].

2.7 PREVIOUS STUDY

This research was conducted to determine the characterization of nickel cobalt oxide as a high-cycle supercapacitor with a variety of solvents, to differentiate it from previous studies, and to add insight from existing research journals. The research conducted in [5] discussed the characterization of nickel cobalt oxide as a potential material for supercapacitors using the hydrothermal method. A simple, scalable, and environmentally friendly hydrothermal synthesis technique was applied to prepare hexagonal NiCo$_2$O$_4$ nanoparticles. Variations in AC conductivity show Jonscher's power law. Electrical conductivity describes areas such as plateaus at low frequencies and conductivity rises at high frequencies. According to [5], conductivity is very important to obtain high power and energy in supercapacitors. The conductivity of a material can decrease due to defects, impurities, and dislocations. Nickel cobalt oxide (NiCo$_2$O$_4$) is composed of a closed cubic spinel structure packed with cations on both the tetrahedral (A) and octahedral (B) sides so that the exchange of electrically charged cations on the A-B side is very important when compared to the B-B side.

Research [27], titled "Preparation and Electrochemical Properties of Nanostructured Porous Spherical NiCo$_2$O$_4$ materials," discusses spherical porous NiCo$_2$O$_4$ powder with a micro-nano structure made by spray drying method using citric acid as a binder and solvent salts as a source of cobalt and nickel. Micro-nano structured porous NiCo$_2$O$_4$ microspheres prepared by the spray drying method exhibit superior electrochemical properties; spherical porous NiCo$_2$O$_4$ powders with micro-nano structure are promising candidates for supercapacitors. The spherical porous NiCo$_2$O$_4$ electrode material with excellent micro-nano structure exhibits characteristics of pseudocapacitor, diffusion, and charge transfer, and enhancement of structure stability and life cycle during the charge-discharge process.

2.8 RESULT AND DISCUSSION

2.8.1 PHASE IDENTIFICATION

The XRD chart in Figure 2.14 shows that all samples are single-phase $NiCo_2O_4$ with peak crystallites of [311], [400], and [440] in samples of 400°C variation; the highest peak is [400]. The sample with 60 minutes variation has a crystallite of 8.977 nm, the sample with 120 minutes duration has a crystallite of 6.279 nm, and the sample with 180 minutes duration has a crystallite of 7.828 nm, as displayed in Table 2.1.

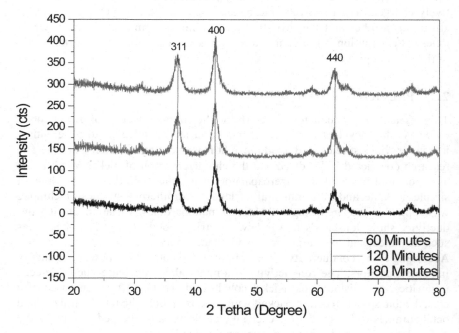

FIGURE 2.14   XRD graph for $NiCo_2O_4$ at sintering duration at 400°C.

TABLE 2.1
Phase Identification of $NiCo_2O_4$ at 400°C

Sample [Minutes]	Height [cts]	FWHM [°2Th.]	d-Spacing [Å]	D (nm)
60	88.35	0.5510	2.0970	6.69
120	101.60	0.7872	2.0943	4.68
180	93.02	0.6298	2.0916	5.88

Figure 2.15 presents that all samples with 500°C temperature variation form single-phase with each having five peaks: [220], [311], [400], [511], and [440], with the highest being [400]. The sample with 60 minutes variation has a crystallite of 24.712 nm, the sample with 120 minutes duration has a crystallite of 19.151 nm, and the sample with 180°C minutes variation has 28.874 nm crystallite, as shown in Table 2.2. Figure 2.16 presents a similar matter to Figure 2.15, where samples have a single-phase NiCo₂O₄ with five peaks: [220], [311], [400], [511],

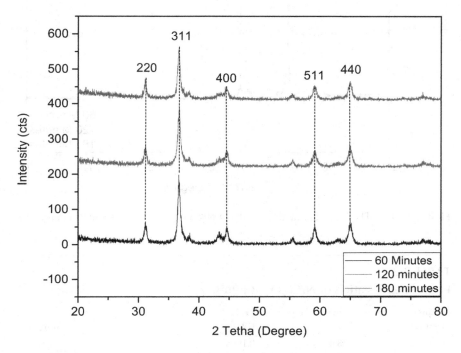

FIGURE 2.15 The X-ray diffraction results of $NiCo_2O_4$ with various sintering duration at 500°C.

TABLE 2.2
Phase Identification of NiCo₂O₄ at 500°C

Sample [Minute]	Height [cts]	FWHM [°2Th.]	d-Spacing [Å]	D (nm)
60	173.7	0.2755	2.4496	24.71
120	150.9	0.3542	2.4453	19.151
180	141.1	0.2362	2.4518	28.874

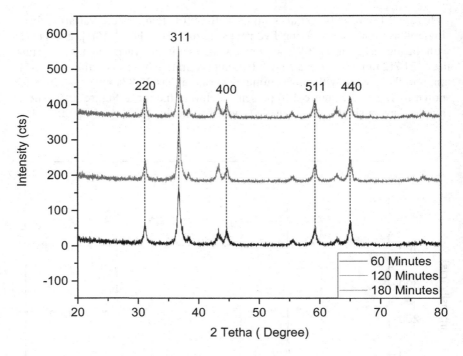

FIGURE 2.16   The X-ray diffraction results of $NiCo_2O_4$ with various sintering duration at 600°C.

TABLE 2.3

Phase Identification of $NiCo_2O_4$ at 600°C

Sample [Minute]	Height [cts]	FWHM [°2Th.]	d-Spacing [Å]	D (nm)
60	148.46	0.1968	2.44907	34.58
120	156.19	0.3542	2.44670	19.17
180	188.68	0.1968	2.44586	34.48

and [440], with [400] as the highest peak. The sample with 60 minutes variation has a crystallite of 34.577 nm, the sample with 120 minutes duration has a crystallite of 19.172 nm, and the sample with 180°C minutes variation has 34.485 nm crystallite, as shown in Table 2.3. Based on these three charts, the $NiCo_2O_4$ nanoparticle has a cubic structure and is centered on the surface [11]–[15], supported by the images from the SEM test. From all charts, the high- and low-intensity peaks display the size of the crystallite, and the crystallite particle can be calculated using the Scherrer formula [16], [17].

2.8.2 MORPHOLOGICAL ANALYSIS

Based on the Figure 2.17, generally, all nine samples have similarities in their morphologies. The nanopowder morphology was close and adhered to each other and form nano bulks. The content from the EDAX test was closer to the theory, in which nickel (Ni) was higher than cobalt (Co) with the ratio of 2:1. However, differing from the theory, the oxygen level here was high due to impurities that causes an increase in the oxide level. Moreover, there were large-size items as a result of particle unification during the sintering process[18], [19]. All the sample's images show that the particle sizes are nano, relatively consistent with the XRD method that applies the Scherrer formula [16], [17]. The increase in sintering temperature was also followed with the increase of particle size, which also meant the surface of other particles, as impurities disappear [19]–[23].

2.8.3 MAGNETIC PROPERTIES

The VSM test results in Figure 2.18 and Table 2.4 show that at 400°C, the highest saturation was found in the sample with the 60 minutes variant, with the value of 1.343 emu/g; whereas the highest retentivity was found in the sample with the 120 minutes variant, with the value of 0.140 emu/g; and the highest coercivity was found in the sample with the 180 minutes variant, with the value of 0.0308 T. The samples at 500°C (Figure 2.19 and Table 2.5) had the highest saturation and coercivity in the sample with 180 minutes duration with the value of 4.781 emu/g and 0.0240 T and highest retentivity in the sample with 60 minutes duration with the value of 1.235 emu/g. The samples at 600°C (Figure 2.20 and Table 2.6) had the all the highest saturation, retentivity, and coercivity in the sample with 60 minutes duration, with the value of 1.986 emu/g, 0.215 emu/g, and 0.0240 T. It can be observed that from all results that the highest saturation and retentivity occurred in the sample with 500°C temperature variant and 180 minutes and 60 minutes duration with the value of 4.781 emu/g and 1.235 emu/g, respectively, while the highest coercivity occurred in the sample with 400°C temperature and 180 minutes duration that was 0.0308 T.

Figure 2.18 displays the curve characteristic where variants with 400°C temperature have superparamagnetic properties, supported by the small particles at 7 ± 2 nm and the highest coercivity value of 0.0308 T [24]. Meanwhile, variants with 500°C have higher saturation and retentivity with the value of 4.781 emu/g and 0.0240 T but lower coercivity with the value of 0.0240 emu/g. This event was caused by the half NiO phase that turned into antiferromagnetic [25], [28]. Variants at 600°C also experienced a decrease in coercivity, saturation, and retentivity. One cause of the decrease was the temperature increase that, in turn, caused the NiO phase from ferromagnetic into antiferromagnetic and reduced the NiO phase [25], [29]–[31]. See Figure 2.21 and Table 2.7.

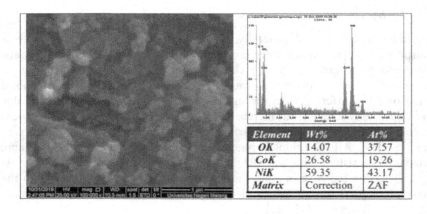

Element	Wt%	At%
OK	14.07	37.57
CoK	26.58	19.26
NiK	59.35	43.17
Matrix	Correction	ZAF

(a)

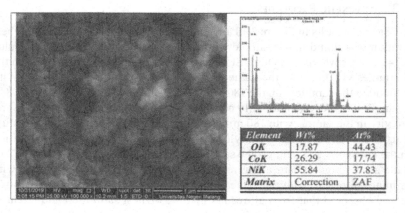

Element	Wt%	At%
OK	17.87	44.43
CoK	26.29	17.74
NiK	55.84	37.83
Matrix	Correction	ZAF

(b)

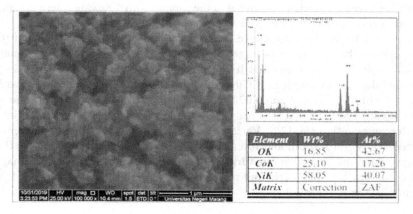

Element	Wt%	At%
OK	16.85	42.67
CoK	25.10	17.26
NiK	58.05	40.07
Matrix	Correction	ZAF

FIGURE 2.17 Morphologies of $NiCo_2O_4$ at 400°C with sintering duration of (a) 60 minutes, (b) 120 minutes, and (c) 180 minutes.

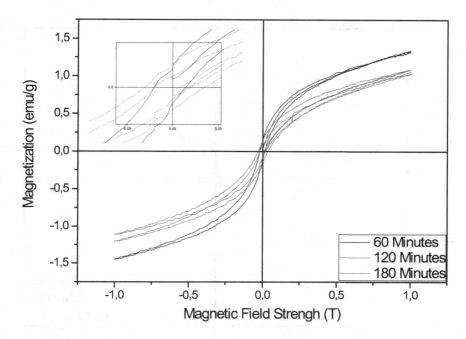

FIGURE 2.18 Hysteresis curve of NiCo$_2$O$_4$ with different sintering times at 400°C temperature.

TABLE 2.4
Magnetic Property Analysis of NiCo$_2$O$_4$ Based on Hysteresis Curve at 400°C Temperature

Sample [minute]	Hc [T]	Mr [emu/g]	Ms [emu/g]
60	0.0202	0.128	1.343
120	0.0201	0.140	1.056
180	0.0308	0.072	1.086

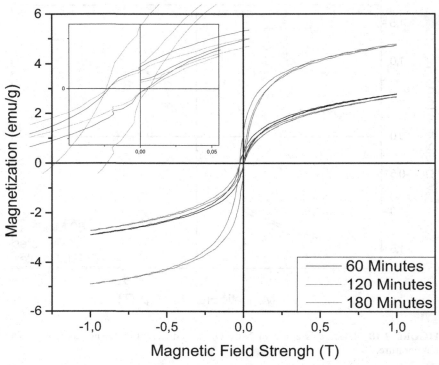

FIGURE 2.19 Curves comparison of $NiCo_2O_4$ with different sintering times at 500°C temperature.

TABLE 2.5

Magnetic Properties of $NiCo_2O_4$ Based on Hysteresis Curve at 500°C Temperature

Sample [minutes]	Hc [T]	Mr [emu/g]	Ms [emu/g]
60	0.0219	1.235	2.781
120	0.0227	0.438	2.669
180	0.0240	0.352	4.781

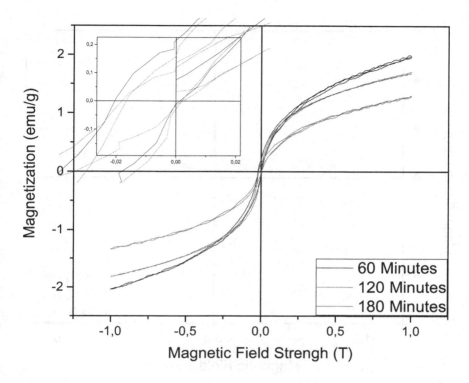

FIGURE 2.20 Curves comparison of NiCo$_2$O$_4$ with different sintering times at 600°C temperature.

TABLE 2.6

Magnetic Property Analysis of NiCo$_2$O$_4$ Based on Hysteresis Curve at 600°C Temperature

Sample [minute]	Hc [T]	Mr [emu/g]	Ms [emu/g]
60	0.0204	0.215	1.986
120	0.0180	0.119	1.301
180	0.0166	0.162	1.706

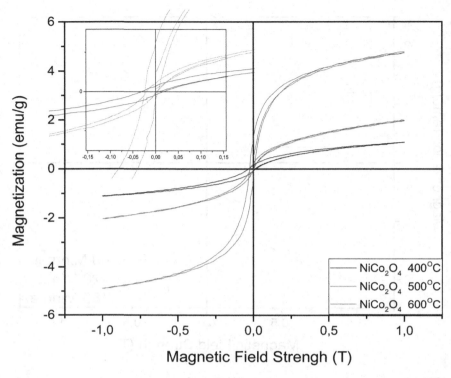

FIGURE 2.21 Hysteresis curves of $NiCo_2O_4$ with different sintering temperatures at 400°C, 500°C, and 600°C temperature.

TABLE 2.7

Magnetic Properties Analysis of $NiCo_2O_4$ Based on Hysteresis Curve at 400°C, 500°C, and 600°C Temperature

Sample [minute]	Hc [T]	Mr [emu/g]	Ms [emu/g]
400°C	0.0308	0.072	1.086
500°C	0.0240	0.352	4.781
600°C	0.0204	0.215	1.986

2.9 CONCLUSION

Synthesis and characterization using the self-combustion method were successfully conducted. The phase identifications showed that the samples had a single phase, as proven by the matching peak of NiCo$_2$O$_4$. All charts with sintering temperature variations displayed the NiCo$_2$O$_4$ nanoparticle in a cubic structure centered on the surface, supported by the morphological analysis. The morphological analysis of NiCo$_2$O$_4$ presented the conversion from a small particle becoming a larger particle as the sintering temperature increases. The shape of NiCo$_2$O$_4$ shows as a nanosphere powder, while the EDAX test resulted in the domination of Ni elements. The magnetic properties from the VSM test showed the highest coercivity at 400°C temperature and the highest retention at 500°C temperature, while samples at 600°C had a declination of coercivity, retentivity, and saturation. Samples with higher temperatures had lower coercivity because of the decrease or disappearance of the NiO phase from ferromagnetic into antiferromagnetic. Therefore, the best material for supercapacitors is NiCo$_2$O$_4$ sintered at 500°C for 60 minutes.

REFERENCES

[1] Badan Pusat Statistik, 'Kebutuhan Energi Indonesia Diproyeksikan Capai 2, 9 Miliar Setara Barel Minyak pada 2050 | Databoks', November, 2021. https://databoks.katadata.co.id/datapublish/2021/12/03/kebutuhan-energi-indonesia-diproyeksikan-capai-29-miliar-setara-barel-minyak-pada-2050

[2] J. Giwangkara, 'The Urgency of Renewable Energy Transition in Indonesia', *Problema Transisi Enegi di Indonesia*, pp. 1–24, 2021.

[3] S. D. Dhas et al., 'Synthesis of NiO Nanoparticles for Supercapacitor Application as an Efficient Electrode Material', *Vacuum*, vol. 181, no. July, p. 109646, 2020, doi:10.1016/j.vacuum.2020.109646.

[4] T. Usami, S. A. Salman, K. Kuroda, M. K. Gouda, A. Mahdy, and M. Okido, 'Synthesis of Cobalt-Nickel Nanoparticles via a Liquid-Phase Reduction Process', *Journal of Nanotechnology*, vol. 2021, 2021, doi:10.1155/2021/9401024.

[5] Poonam, 'Characterization of Nickel Cobalt Oxide: A Potential Material for Supercapacitor', *Nanotechnology*, vol. 29, no. 27, 2018.

[6] R. Kumar, *NiCo2O4 Nano-/Microstructures as High-Performance Biosensors: A Review*, vol. 12, no. 1. Springer, Singapore, 2020, doi:10.1007/s40820-020-00462-w.

[7] S. Patil and H. P. Dasari, 'Effect of Fuel and Solvent on Soot Oxidation Activity of Ceria Nanoparticles Synthesized by Solution Combustion Method', *Materials Science for Energy Technologies*, vol. 2, no. 3, pp. 485–489, 2019, doi:10.1016/j.mset.2019.05.005.

[8] M. S. Alshatwi, H. A. Alburaih, S. S. Alghamdi, D. A. Alfadhil, J. A. Alshehri, and F. A. Aljamaan, 'Iron-doped Nickel Oxide Nanoparticles Synthesis and Analyzing Different Properties', *Advances in Science, Technology and Engineering Systems Journal*, vol. 6, no. 1, pp. 1422–1426, 2021, doi:10.25046/aj0601161.

[9] Sani Garba Danjumma, 'Nickel Oxide (NiO) Devices and Applications: A Review', *International Journal of Engineering Research and*, vol. 8, no. 4, 2019, doi:10.17577/ijertv8is040281.

[10] R. Palombari, 'Influence of Surface Acceptor-donor Couples on Conductivity and Other Electrochemical Properties of Nonstoichiometric NiO at 200°C', *Journal of Electroanalytical Chemistry*, vol. 546, no. SUPP, pp. 23–28, 2003, doi:10.1016/S0022-0728(03)00134-7.

[11] Z. Alhashem, C. Awada, F. Ahmed, and A. H. Farha, 'Structural and Magnetic Properties Study of Fe2O3/NiO/Ni2FeO4 Nanocomposites', *Crystals*, vol. 11, no. 6, pp. 1–12, 2021, doi:10.3390/cryst11060613.

[12] E. Krüger, 'Wannier States of FCC Symmetry Qualifying Paramagnetic NiO to be a Mott Insulator', *Symmetry*, vol. 12, no. 5, pp. 10–13, 2020, doi:10.3390/SYM12050687.

[13] H. T. Rahal, R. Awad, A. M. Abdel-Gaber, and D. E. S. Bakeer, 'Synthesis, Characterization, and Magnetic Properties of Pure and EDTA-Capped NiO Nanosized Particles', *Journal of Nanomaterials*, vol. 2017, 2017, doi:10.1155/2017/7460323.

[14] P. Puspitasari, D. R. Qomarudin, S. Sukarni, A. A. Permanasari, J. A. Razak, and R. Nurmalasari, 'Synthesis and Characterization of Cobalt Oxide Powder with Sintering Duration Variation by Sol-Gel Method', *Intelligent Manufacturing and Mechatronics*, pp. 549–555, 2022, doi:10.1007/978-981-16-8954-3_53.

[15] J. Pal and P. Chauhan, 'Study of Physical Properties of Cobalt Oxide (Co 3 O 4) Nanocrystals', *Materials Characterization*, vol. 61, no. 5, pp. 575–579, 2010, doi:10.1016/j.matchar.2010.02.017.

[16] J. H. Lin et al., 'S Doped NiCo2O4 Nanosheet Arrays by Ar Plasma: An Efficient and Bifunctional Electrode for Overall Water Splitting', *Journal of Colloid and Interface Science*, vol. 560, pp. 34–39, 2020, doi:10.1016/j.jcis.2019.10.056.

[17] W. Callister, *Materials Science and Engineering An Introduction*, 10th Edition by David G. Rethwisch and William Callister, vol. 53, no. 9, 2017. https://ftp.idu.ac.id/wp-content/uploads/ebook/tdg/TEKNOLOGI%20REKAYASA%20MATERIAL%20PERTAHANAN/Materials%20Science%20and%20Engineering%20An%20Introduction%20by%20William%20D.%20Callister,%20Jr.,%20David%20G.%20Rethwish%20(z-lib.org).pdf

[18] J. A. Buck, *Engineering Electromagnetics*, 9th edition, 2017.

[19] R. S. Rawat and Y. Wang, *Plasma Nanotechnology for Nanophase Magnetic Material Synthesis*, 2020. doi:10.4324/9781315371573-4.

[20] N. Baig, I. Kammakakam, W. Falath, and I. Kammakakam, 'Nanomaterials: A Review of Synthesis Methods, Properties, Recent Progress, and Challenges', *Materials Advances*, vol. 2, no. 6, pp. 1821–1871, 2021, doi:10.1039/d0ma00807a.

[21] K. W. Prasetiyo, 'Aplikasi Nanoteknologi Dalam Industri Hasil Hutan (Application of Nanotechnology in Forest Products Industry)', *Jurnal Akar*, vol. 9, no. 1, pp. 13–24, 2020, doi:10.36985/jar.v9i1.189.

[22] S. Bayda, M. Adeel, T. Tuccinardi, M. Cordani, and F. Rizzolio, 'The History of Nanoscience and Nanotechnology: From Chemical-physical Applications to Nanomedicine', *Molecules*, vol. 25, no. 1, pp. 1–15, 2020, doi:10.3390/molecules25010112.

[23] J. Mohanta, B. Dey, and S. Dey, 'Magnetic Cobalt Oxide Nanoparticles: Sucrose-Assisted Self-Sustained Combustion Synthesis, Characterization, and Efficient Removal of Malachite Green from Water', *Effective Science Communication*, vol. 65, no. 5, pp. 2819–2829, 2020, doi:10.1021/acs.jced.0c00131.

[24] E. Carlos, R. Martins, E. Fortunato, and R. Branquinho, 'Solution Combustion Synthesis: Towards a Sustainable Approach for Metal Oxides', *Chemistry—A European Journal*, vol. 26, no. 42, pp. 9099–9125, 2020, doi:10.1002/chem. 202000678.

[25] X. U. Jing, C. Qinghua, L. Xinping, and H. Baoquan, 'The Combustion Synthesis of Nanoscale Ceria-doped Calcia Solid Solution', *Advanced Materials Research*, vol. 613, pp. 547–550, 2013, doi:10.4028/www.scientific.net/AMR.610-613.547.

[26] M. S. M. Basri, N. Mazlan, and F. Mustapha, 'Effects of Stirring Speed and Time on Water Absorption Performance of Silica Aerogel/Epoxy Nanocomposite', *Journal of Engineering and Applied Sciences*, vol. 10, no. 21. pp. 9982–9991, 2020, doi:10.1039/d0ra00259c.

[27] C. Li, Y. Liu, G. Li, and R. Ren, 'Preparation and Electrochemical Properties of Nanostructured Porous Spherical NiCo2O4 Materials', *RSC Advances*, vol. 10, no. 16, pp. 9438–9443, 2020, doi:10.1039/d0ra00259c.

[28] P. Puspitasari, *X-Ray Diffraction: Teori Dasar, Perhitungan dan Solusi Permasalahan*. UM Press, 2019.

[29] Y. Leng, *Materials Characterization: Introduction to Microscopic and Spectroscopic Methods*. Wiley, 2019.

[30] Z. Sherif, G. Xian, S. Thomas, and A. Ajith, 'Effects of Surface Grafting of Copper Nanoparticles on the Tensile and Bonding Properties of Flax Fibers', *Science and Engineering of Composite Materials*, vol. 24, no. 5, pp. 651–660, 2017, doi:10.1515/secm-2014-0462.

[31] S. Kumar, *Handbook of Materials Characterization*. Springer, 2018, doi:10.1007/978-3-319-92955-2_3.

3 Variety of Agricultural Machine Innovations by Utilizing Renewable Energy

Riana Nurmalasari and Poppy Puspitasari

CONTENTS

3.1 INTRODUCTION

One of the significant industries that affect the economies of all nations around the world is the agricultural sector. This is due to the fact that the agricultural industry plays a significant role in maintaining the supply of essential goods in every nation. Agriculture is the practice of utilizing natural resources by human beings to produce food, industrial raw materials, or energy sources, as well as to manage the environment. This activity is also known as environmental management. Agriculture can be defined in two ways: broadly, as all activities that involve the use of living things (including plants, animals, and microbes) for the benefit of humans, and narrowly, as the activity of using a piece of land to

DOI: 10.1201/9781003367819-3

cultivate specific types of plants. Both definitions of agriculture are used inter-changeably throughout this article. In addition, the agricultural sector contributes to equitable development through its contributions to the fight against poverty and the improvement of people's incomes. In addition, the agricultural industry has emerged as a significant force in the formation of the nation's culture as well as the maintenance of ecological equilibrium.

Since the dawn of civilization, people have lived and worked in the agricul-tural industry, even a very long time ago, before the beginning of modern human civilization. The agricultural industry is experiencing rapid expansion as a direct result of current events. One of these advantages is the incorporation of various forms of technology into today's agricultural practices. In today's agriculture, the emphasis is placed on productivity, efficiency, quality, and maintenance of a continuous supply of food, all of which need to be continually increased and maintained. In order to compete in the global market, agricultural goods, such as food crops (horticulture), commodities, fisheries, plantations, and livestock, must be packaged to a high standard and quality. These high-quality products are, without a doubt, manufactured through the application of technological con-tent at some stage of the production process.

A number of nations are vying with one another to develop agricultural tech-nology. There is a connection between one of them and the equipment that is utilized in agricultural settings. Unfortunately, in some countries, not all tech-nologies can be adopted and utilized in quite the same way as described earlier. This is due to the fact that agriculture in a number of different nations possesses different characteristics, and even the conditions of agricultural land in each region possess their own unique characteristics. Because of this, the advance-ment of technology needs to be adapted to the agricultural conditions that are present in that country. There are a lot of things that play a role, such as the sea-sons, land conditions, natural resources, and risks of natural disasters.

The innovations that are related to technological development in agriculture have the potential to continue to be developed in a more comprehensive manner (Rehman et al., 2016). Utilizing sources of energy that do not deplete Earth's resources is one option. At this time, agricultural machinery is still heavily reli-ant on energy sources that are not renewable. Therefore, there is a need for addi-tional research and development concerning technological innovations that make use of renewable energy for agricultural purposes.

3.2 DEVELOPMENT OF AGRICULTURAL TECHNOLOGY

The field of agriculture is one that plays a significant part in the human life cycle. The reason is that this is the foundation upon which we build our ability to pro-vide humans with clothing, food, and a place to live. In addition to this, the agricultural industry serves as the cornerstone upon which people's everyday lives are built. In the realm of agriculture, the role that technology plays in ensur-ing the success of the farming productivity that results from it is indispensable (Manida and Ganeshan, 2021). In addition, as the number of people in the world increases, there will inevitably be greater demand for things like food, clothing, and housing. In particular, the requirement for the consumption of food (Sadiku et al., 2021). Simply put, if people do not have access to food, they will not be able to continue living. There is also the question of whether or not the availability of food in a country is an indication of that country's level of prosperity. Because of this, people in the agricultural industry around the world will need to exert more effort to satisfy these increased demands for food. The process is carried out one step at a time in order to maximize the output of the final product.

In the modern era, information can be obtained quickly and easily through a variety of supporting media, including the internet, television, and print media. In this scenario, the world of agriculture also makes use of information technology to support activities related to the development of sustainable agriculture (Yaseen et al., 2019). The timely realization of modern agriculture is dependent on information and communication technology's ability to play an important role.

At this time, mastery of information technology is becoming increasingly more advanced. At this time, information technology is an unquestionable fact that cannot be debated. Many people believe that information technology is a tool for change that can bring about greater ease in the activities of daily life, which can then lead to a greater number of benefits from information technology. One of the many areas that could benefit greatly from the application of information technology is agriculture. It is anticipated that progress will be made in the agricultural sector through the utilization of effective information technology.

The use of information technology in any form is essential to the modern agricultural industry. Farmers have a need for information that is relevant to the field of agriculture. In this age of information, it is not difficult to acquire this knowledge thanks to the widespread dissemination of media throughout society. Information on the outcomes of research and innovation in the agricultural sector lends a hand to efforts to increase the production of agricultural commodities, which in turn helps to bring about the agricultural development that was anticipated. The dissemination of agricultural information and expertise will serve as a catalyst for the generation of opportunities for agricultural and economic development, ultimately leading to the alleviation of poverty. Information and communication technology helps provide information that is pertinent and timely, which makes it simpler for farmers to make decisions about opportunities and produce the most amount of product possible.

In agriculture, communication is extremely important for the formation of networks among farmers and among agencies that support agricultural development. The problem of production of agricultural commodities, if there are problems or a need to supply, is now no longer a problem as a result of the communication that exists between farmers in other areas. This is in order to empower farmers to make the most informed decisions possible regarding the management of their agricultural lands. In a similar vein, with regard to other challenges, they can be conquered by conversing with one another.

Farmers need access to a wide range of information in the agricultural sector in order to effectively manage their farming businesses. This information

includes government policies, research results from a variety of fields, the expe-
riences of other farmers, and up-to-date information regarding market prospects
related to production facilities and agricultural products. They have access to
a variety of sources of information, and one of those sources is the internet.
Farmers who have access to the internet can obtain a variety of information per-
taining to agriculture. In addition to this, they have the opportunity to obtain the
most recent information concerning the prospects of the international market in
relation to production inputs and agricultural products. However, the government
is also obligated to assist farmers in producing high-quality agricultural goods
by providing them with counseling services.

In addition, information technology plays a part in the marketing of agricultural products. Many different kinds of businesses are becoming more adaptable
to the advances that have been made in information technology. Conventional
methods of conducting business are not utilized nearly as frequently as newer
methods, which involve marketing products in cyberspace. These newer methods include marketing through web media, conducting transactions online, and
even marketing through social networks. The marketing of agricultural products
through the internet is unquestionably more cost-effective than traditional methods. Farmers are able to easily learn what consumers want. Farmers are able to
coordinate their planting efforts to ensure that there will always be a sufficient
supply on the market, that prices will remain consistent, and that availability will
not fluctuate. Farmers can sell their agricultural products more quickly if they are
able to communicate more quickly.

As long as people continue to have a need for food, there will always be a
market for agricultural products and demand will only grow as the world's population does. There are already many farmers in developed countries, such as the
United States, Japan, and Korea, who market their agricultural products via the
internet. Other examples of these countries include the United Kingdom. They
are able to monitor marketing by using a website that was developed specifically
for the process of buying and selling.

The application of information and communication technology in farming on
a large scale is a process that can significantly benefit from its application. A
good illustration of this would be the application of a program to assist in the
mapping of agricultural plots. Farmers are able to more easily estimate or ana-
lyze the results of their agricultural production as well as determine the direction of pollination and the spread of disease, among other things, thanks to this
method. Another illustration of this would be the utilization of an airplane as part
of the process of fertilization on a large and extensive scale; the airplane would
be equipped with an automated fertilizer spraying machine, and the airplane
would be designed according to the requirements. Farmers can gather information about the condition of their agricultural land without having to physically
travel to the field because the aircraft is also outfitted with a camera sensor that
can see the land's condition. One of the many positive effects that have resulted
from the progression of information and communication technology in this world
is the illustration presented here.

3.3 PEST REPELLENT MACHINE FOR AGRICULTURE

Numerous pests and diseases attack rice plants, resulting in a potential decrease in rice production. In remote regions of Indonesia, there are still numerous farmers who are unaware of the pests and diseases that attack rice plants. Moreover, they do not know how to effectively control the pests and diseases that attack rice plants. As a result of crop failure, there are fluctuations in the rice cultivation industry. Crop failure has a significant impact on farmers as well as on the greater community, as the quantity of essential food ingredients is diminished (Adhitya, 2018). This failure occurred as a result of pests and diseases attacking the rice plants. There are numerous potential diseases and pests that can attack rice plants. Grasshoppers and leaf rust are two examples.

Pests and diseases in rice plants impede the growth and development of the crop (Syahminan, 2017). As for the diverse pests and diseases that attack rice plants, these include planthoppers, walang sangit, birds, rats, golden snails, and ground bedbugs, as well as brown spot disease, blast, leaf blight, tungro, leaf brown line disease, leaf midrib rot, fusarium disease, false fire, leaf streak, and dwarf disease. These pests and diseases can have a negative impact on the growth and yield of rice plants. Therefore, a variety of control methods are required.

In almost every country in the world, farmers face difficulties caused by various kinds of pests. This is due to the fact that one of the factors that affect crop yields is the presence of pests. There have been a lot of efforts put into reducing the damage that pest infestations cause. Beginning with the pharmaceutical industry's use of chemical drugs, natural medicines, and various forms of media used to repel pests. Every measure taken to lessen the damage caused by pest infestations has some unintended consequences. If chemical drugs are used for an extended time, they have a negative impact on the environment and lead to the production of non-organic yields. The effects of natural remedies don't tend to last very long, particularly if the pests have developed immunity to them. In addition, the application of pest repellents can frequently involve a level of risk that is fairly high. Take for instance the application of electric wires as a rodent deterrent. Electrocution was the cause of death in a significant number of farming accidents. Additionally, there is a great deal of additional effects that are brought about by the utilization of midges, particularly those that are still conventional.

The field of agriculture is continuing to see new developments in technology related to midges up until the present day (Nair et al., 2017; Onah and Iloka, 2013). One of these options is a device that employs solar power as its primary source of power in order to combat pests (Figure 3.1). The shape of this implement is meant to resemble a scarecrow so that the birds that typically eat rice will believe that a farmer is standing guard over the crop and avoiding eating it. The solar panels that are affixed to the top of the scarecrow will collect thermal energy from the sun during the day and store it so that the insect repellent will continue to function even when the sun goes down. The component of the system that has been installed in this instrument takes the form of a sound sensor that reflects specific waves. Birds, mice, and other insects have a strong aversion to these sound waves

FIGURE 3.1   Pest repellent using solar power.

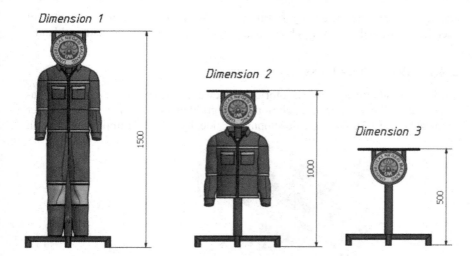

FIGURE 3.2 Adjustable dimension for pest repellent using solar power.

(Aflitto and DeGomes, 2020). Therefore, if this tool is activated, it will significantly reduce the number of pests that approach agricultural land and cause damage to it. This instrument also includes a frequency wave that can encourage the
growth of plants and is equipped with this feature. Additionally, this instrument
is fitted with day and night sensors for optimal performance.

The height of pest repellent, as shown in Figure 3.2, can be adjusted according
to need. This makes it simpler for farmers to use it according to the requirements
of the land and the conditions of agricultural land. Rice-eating birds can be scared
away if the size of the barrier is 1.5 meters. Pest repellent can grow to a length of
1.5 meters and have the appearance of scarecrows. When used at a size of 1 meter,
it is effective at warding off insects and other pests of a smaller size. The shape of
the legs is removed when the scarecrow is 1 meter in height, leaving only the upper
body of the scarecrow. In addition, when it is 0.5 meters in length, it can be utilized
to ward off mice and can promote the growth of plants. When they have grown to a
size of 0.5 meters, the pest repellent does not take the form of scarecrows.

3.4 AGRICULTURAL WATER SOURCES
USING RENEWABLE ENERGY

Agriculture relies heavily on the availability of water (Hans, 2018). This is due
to the fact that plants require water in order to thrive (Zarate et al., 2014). The
best-case scenario for farmers would be if agricultural land was blessed with an
abundance of springs. In most cases, farmers will use wells as a method to disperse
water across the agricultural land that they own. In addition to that, some farmers
also irrigate their land using pumps. A motor that runs on either gasoline or diesel

fuel is what provides the energy for turning on the pump. Therefore, the amount of fuel that is utilized determines how much it will cost to irrigate agricultural land.

3.4.1 WATER PUMP USING WIND POWER

Utilizing wind power, which is a form of renewable energy, is one of the innovations that has been made as a solution to the problem of the use of fuel for water pumps (Figure 3.3). This new development is ideally suited for implementation in

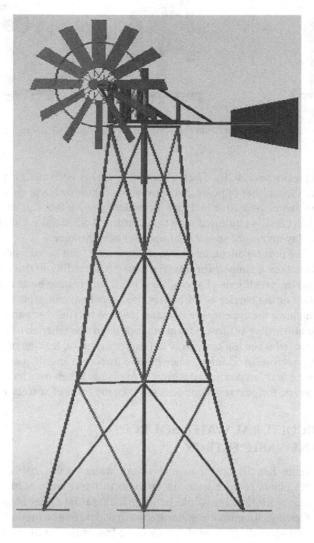

FIGURE 3.3 Design of water pump using wind power.

FIGURE 3.4 Implementation of water pump using wind power.

agricultural land irrigation efforts in regions that experience wind throughout the entire year (Figure 3.4). Towers atop where windmills are mounted can be used in the construction of water pumps that are powered by wind energy. A wind direction regulator is built into the windmill so that it can continue to rotate in response to the prevailing wind in any given direction. After that, the energy is transferred from the rotating mill to the drive so that it can continuously operate the pump. By going through this process, water from the ground can be brought up to irrigate land that is used for farming.

3.4.2 WATER PUMP USING EARTH GRAVITY

Another innovation for the irrigation process is the utilization of Earth gravity–based pumps (Figure 3.5). This is in addition to the use of wind energy, which is also one of the innovations. The operation of this pump does not require the use of electricity. In order for the pump to function properly and distribute water, there must be a minimum distance of one meter between it and the source of the water, as well as an input pipe that is at least 8 meters in length. The water source should be located as high up as possible. In a similar vein, the length of the input pipe should be increased whenever possible. This pump raises the water to a higher level by utilizing the pressure that is created by the height difference in addition to the hydraulic pressure that is already present.

For the purpose of irrigation, this technology can be utilized to channel water from springs that are situated in remote valleys or even farther away. This pump operates without the need for any electricity or fuel of any kind. This pump has the capability of raising water to a vertical height of between 50 and 80 meters, and it can transport water up to a distance of 2 kilometers when conditions are optimal. The water flow rate is 30 liters per minute at the head of 30 to 50 meters.

FIGURE 3.5 Implementation of water pump using Earth gravity.

3.5   FERTILIZER SPREADER MACHINE FOR AGRICULTURE

Utilizing fertilizers is one of the supplementary practices that are utilized in order to achieve maximum agricultural yields. It is possible to improve the fertility and overall health of plants by applying fertilizer (Savei, 2012; Veronica et al., 2015). In most cases, farmers will either utilize chemical fertilizers or natural fertilizers in their crop production (Mahdi et al., 2010). There are a few different approaches that can be taken when it comes to the process of applying fertilizer; among these approaches, the conventional method is one that is still utilized quite frequently (Alhassan et al., 2019). Mainly due to the fact that the traditional method of applying fertilizer takes a considerable amount of time, particularly if the area being worked is particularly large.

The use of an automatic solar fertilizer spreader machine is one of the innovations that have been developed to simplify the process of fertilizing agricultural land (Figure 3.6). Arduino was used to write the code that controls this

FIGURE 3.6   Solar fertilizer spreader machine.

automatic fertilizer-spreading machine, which reads data from an ultrasonic sensor. Batteries are used to store the solar energy that is produced by solar panels, which is then used to power devices, such as electric motors and stepper motors. This machine is driven by an ultrasonic sensor (HC-SR04) that is located on the front of the machine. This sensor will drive the stepper motor that is connected to the control of the front wheels. The automatic control of this machine is carried out by an ultrasonic sensor (HC-SR04). If there is something in the way of the machine, the sensor will detect it and signal the stepper motor to begin turning the steering wheel. This will happen automatically if there is an obstruction in the way.

Figure 3.7 shows that the ultrasonic sensor (HC-SRO4) at number 1 will detect whether there are plants or not in the front machine. If there are plants in front of it at a distance of 5 cm, the ultrasonic sensor will read the distance of the plants and then pass on the information to Arduino, which is number 2. In a nutshell, the working principle of this fertilizer spreader machine is as follows: the ultrasonic sensor will detect whether there are plants or not in front of it. The information will first be processed by the Arduino, and then it will be sent to the

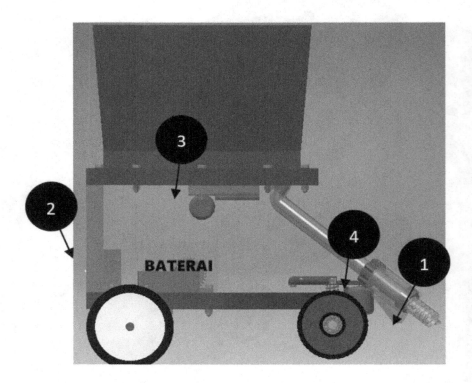

FIGURE 3.7 How the solar fertilizer spreader machine works.

stepper motor located at number 3 so that it can move. The movement of the stepper motor will cause the rope that is connected to the fertilizer opener at number 4 to be pulled, which will result in the fertilizer being released. In order to set the amount of fertilizer, the opening valve number 4 needs to be adjusted.

3.5.1 AGRICULTURE ORGANIC FERTILIZER PROCESSING MACHINES

Farmers typically make use of both chemical fertilizers and organic fertilizers due to the significance of using fertilizers in agricultural practices (Kumar et al., 2013). Some farmers rely solely on chemical fertilizers because they believe them to be the more convenient and productive option. There are farmers who use both chemical and organic fertilizers in their crop rotation. However, farmers who put an emphasis on growing organic crops typically only apply natural fertilizers.

Organic fertilizers are made from natural ingredients. Animal waste, plants that have been processed into compost, and other materials that do not contain any chemical components are common examples of the types of materials that are frequently utilized. There have been many developments in machinery that make use of alternative forms of power in recent years, and one of those developments is the production of organic fertilizers.

A chopping machine that runs on solar power is one of the machines that can be used to make environmentally friendly compost (Figure 3.8). Solar panels are used to collect energy, which is then stored in batteries before being used to power the motor. For the purpose of the composting process, organic waste in the form of green plants is fragmented into smaller pieces whenever the motor moves. Compost is the product of the composting process, and it is this compost that can be applied as a natural fertilizer to plants. The motor that drives this chopper does not require the use of fuel in order to generate the necessary amount of power.

A machine that grinds goat manure and is powered by the sun is another tool that can be used to make organic fertilizer. Compost is made from the residue that is left over after goat manure has been ground up. Composting food scraps produces a by-product that can be utilized as an organic fertilizer. Because this device runs on solar power, there is no need to add fuel to the motor in order for it to function properly. See the solar panels in Figure 3.9, number 2. The utilization of solar energy is planned to lessen people's reliance on various forms of fuel.

It is anticipated that the production of this machine will reduce the amount of chemical fertilizers that are utilized, which, when used repeatedly in rice fields, can compromise the structure of the soil. In addition, it is anticipated that the utilization of this machine will result in a reduction of the costs associated with plant maintenance in agricultural settings, which are typically brought about by the high cost of fertilizers that are imported from other countries. The utilization of this machine has the potential to lessen the amount of waste produced by livestock manure, particularly goat manure, which is notoriously difficult to decompose due to the relative rigidity of its composition. The application of this machine can also be used for conventional fields to improve the welfare of

FIGURE 3.8   Chopper machine using solar power.

farmers through local farmer organizations or groups, specifically by utilizing the product in the form of compost and selling it. This can be accomplished by using local farmer organizations or groups.

In addition, machines that are still related to the process of processing animal manure so that it can be used as compost material are machines that dry chicken manure, as shown in Figure 3.10. In order to prepare chicken manure for the composting process, this machine is utilized to dry the manure. The utilization of an Arduino Uno as the central processing unit (CPU) in this apparatus represents a significant technological advancement. In order to achieve the best possible

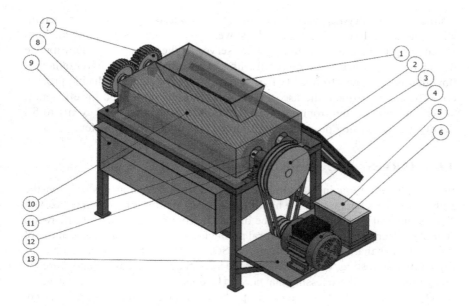

FIGURE 3.9 Grinder machine using solar power.

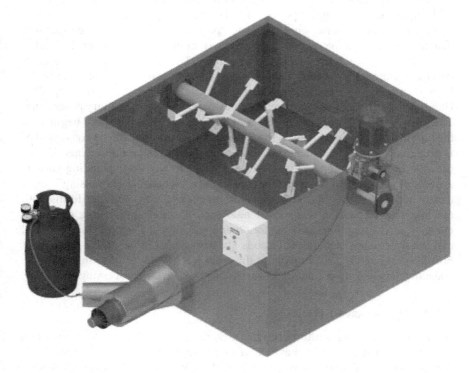

FIGURE 3.10 Dryer machine using Arduino.

results from the drying process, this dryer was built using a bed dryer system, which has a bed-like arrangement. The stove, in its capacity as a heating medium for cooking using a hot air flow system, serves as the primary medium for the drying process. The dryer is outfitted with a blower that acts as a driver of hot air flow that is generated from LPG fuel. This was done so that the drying process could be automated and the use of materials could be optimized for efficiency. Where the hot air flow from the stove with this blower can spread evenly to fill the drying room using the principle of forced convection heat transfer.

3.6   HARVEST PROCESSING MACHINE

In agriculture, producing maximum yields requires focusing not only on the planting process but also on the handling of the crop after it has been harvested. This is in addition to the importance of the planting process. Because there are so many different types of agricultural products, there are also a great number of different ways that they must be handled. Some agricultural products need to be dried after harvesting, while others need to be packaged right away, and there are many other types of processing that need to be done in order to keep the harvest's quality intact. When the harvest is properly cared for after it has been gathered, the quality of the product can be preserved for an extended time. In addition, taking into consideration how the harvest will be handled after it has been gathered will help achieve the highest possible level of harvest quality.

In the context of agriculture, the term "post-harvest" refers to the processes that are carried out on agricultural products after they have been harvested but before they reach the hands of consumers. The term that more accurately describes what happens after harvesting is called "post-production," and it can be broken down into two distinct stages or parts: post-harvest and processing. Handling what occurs after harvest is typically referred to as primary processing. The term "primary processing" refers to any and all treatments that begin with the harvest and continue on until the commodity in question can be consumed or is prepared for the subsequent processing step. This treatment, in general, does not alter the form or appearance of the substance in any way, including various aspects of marketing and distribution. The term "secondary processing" refers to any action that modifies crop yields to other conditions or other forms with the intention of making them last longer (preservation), preventing unwanted changes, or making them suitable for other applications. The processing of food and the processing of industrial goods are both included here. The purpose of post-harvest handling is to maintain the crop in a healthy state so that it can be consumed right away or used as a raw material for further processing.

3.6.1   POST-HARVEST HANDLING

In the handling of agricultural products, there are a number of steps that must be taken immediately after harvest. If these steps are not taken immediately, the quality will degrade and the rate of deterioration will increase, making long-term storage impossible. For crops to produce high-quality yields, proper post-harvest care is required so that it can meet and fulfill the needs of the community. Post-harvest consists of multiple phases, from harvesting to transport to market. Typically, a farmer is unaware of the significance of post-harvest handling and processing, causing a good harvest to deteriorate due to improper handling. Some farmers continue to use traditional tools despite the fact that market demands have become more modern. For instance, when the market requires a fresh mango, farmers frequently harvest mangoes that should not be picked. For this reason, post-harvest treatment tools or technologies that can aid farmers are required.

The water content of the commodity is decreased through the drying process. Grain needs to be dried until it reaches a specific level of moisture content before it can be kept for an extended time. For instance, the drying process for shallots only continues until the skin becomes dry.

The precooling of various fruit and vegetable items: the fruits, once they have been harvested, are immediately stored in a cool or cool place, where they are not exposed to sunlight. This allows the heat that has been carried from the garden to be cooled immediately, which in turn reduces the amount of evaporation that occurs so that the fruit's crispness can be preserved for a longer time. If the facilities are available, the precooling process should be completed in one to two hours at a temperature of approximately 10°C.

Preparation of yams, tubers, and rhizomes for curing. Recovery of shallots, ginger, and potatoes is accomplished by drying them in the sun for one to two hours, up until the point where the soil that was attached to the tubers is dry and can be easily removed and cleaned. After that, it was immediately placed in a location that was cool, dry, or both. Regarding potatoes that are being kept in a dark place right away (no light). Curing is also important for closing wounds that may have been caused during the harvesting process.

It is common practice to bunch leaf vegetables, root tubers (like carrots), and fruits with stems like longans. Binding is done to make handling easier and to reduce the risk of damage.

Leafy vegetables that grow close to the ground are typically washed to remove any dirt that has adhered to their leaves and to maintain their freshness. In addition to that, washing can lessen the presence of pesticide residues as well as insects that carry diseases. It is recommended to use clean water for washing, and it is highly recommended to include disinfectants in the water used for washing. Both potatoes and sweet potatoes should not be washed before consumption. Because the waxy coating on the surface of the fruit is removed when cucumbers are washed, the fruit loses its ability to keep for an extended time after being harvested. Washing bananas can delay maturity.

Cleaning and polishing, specifically freeing something from dirt or other unnatural elements. During this step, you will remove any unwanted parts of the plant, such as leaves, stems, or roots.

The process of sorting involves separating items that can be sold from those that are not commercially viable. Not suitable for retail sale, particularly in the case of produce that is flawed or infected with unwanted organisms, such as insects or pathogens, in order to prevent the contamination of other, healthier crops.

3.6.2 RICE GRINDER MACHINE

The operation of a rice grinder machine is very straightforward (Figure 3.11). A solar panel serves as this device's primary working component. The conversion of solar energy into electrical energy is the primary function of solar panels. The community needed a solution for grinding rice at their individual homes, so this tool was developed as a possible option. People only need to store rice and grind it when they need it because storing rice for an extended time causes it not in good quality; therefore, people do not need to store rice. A direct current motor serves as the driving force behind this rice grinder machine. Pulleys and v-belts are used

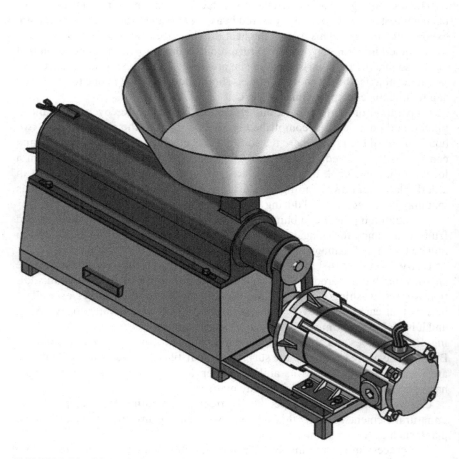

FIGURE 3.11   Rice grinder machine.

to transfer power from the DC motor to the main shaft of the crusher. This allows the crusher to rotate, which is necessary for separating rice from its husks. In order for this machine to function properly, the rice must first be fed into the chimney, where it is then crushed using a crusher. Finally, once the rice has been crushed, it must be filtered in order to separate the rice from the rice bran. After that, the rice will emerge from the machine via one of the doors on the side.

3.6.3 Chopper Machine Using Solar Power

Before the drying and packaging processes are carried out, this chopper's primary purpose is to chop the harvest in the form of medicinal plants into smaller pieces. Solar panels are utilized by this apparatus in order to convert the sun's kinetic energy into electrical energy (Figure 3.12). This tool provides a selection

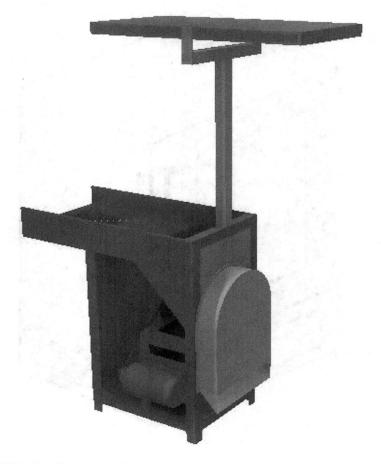

FIGURE 3.12 Chopper machine.

of thicknesses for the results of the chopping, allowing it to be tailored to the requirements of the task at hand. There is a degree of leeway available in terms of the thickness of the chopped results.

3.6.4 Automatic Drying Machine

This dryer makes use of a tray dryer system that is equipped with a blower stove and is controlled by a microcontroller based on the Arduino Uno platform (Figure 3.13). It is anticipated that the use of this machine will improve the process of drying agricultural products in terms of both its effectiveness and its efficiency. The stove, which is used as the heating medium in this apparatus, has

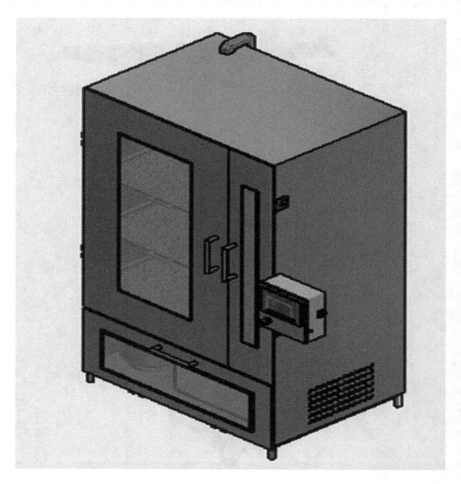

FIGURE 3.13 Automatic drying machine.

been modified by the addition of a blower and a microcontroller-based system. This stove has a blower that is used to push the fire around so that the flames can cover a larger area and the rate at which heat is transferred can be increased. This device is outfitted with a microcontroller component that serves as the working brain of the machine. This component is programmed to command the drying temperature at the appropriate time. The thermocouple is the kind of temperature sensor that's being utilized.

3.6.5 ELECTROPNEUMATIC PEELING MACHINE

This machine can remove the skin and core of pineapples according to its specifications (Figure 3.14). This device's automated movement is achieved through the use of electropneumatics. Once turned on, the majority of the movements

FIGURE 3.14 Electropneumatic peeling machine.

are performed by themselves automatically. After the top of the pineapple and the hump are automatically removed from it with the help of a cutting knife, the pineapple is pushed to the peeler knife so that it can be peeled.

3.7 POSITIVE IMPACTS OF AGRICULTURAL TECHNOLOGY

Farmers need to spend a significant amount of time working the land without the aid of technology in order to achieve extensive land processing. A single day's worth of labor on agricultural land can be cultivated on a three-hectare plot by farmers. The cultivation of farmers' land; however, will be facilitated by the availability of technology, making the process simpler and more efficient; as an illustration, by employing the engine of a tractor. In days gone by, there were no tractors or other machines that could pull harrows; instead, farmers relied on animals like buffalo and cows, or even more primitively, they just used a hoe. Because of this factor, the cultivation of agricultural land is a time-consuming process. In addition to reducing the amount of time needed to complete a task, the application of technology can also have an impact on the amount of output that is obtained from farmers. The variety and quantity of agricultural products being produced both continue to rise. Farmers used to plant regular corn in the past, but nowadays, they can produce hybrid corn by mating different types of plants together. This results in hybrid corn having higher yields and a more aesthetically pleasing appearance.

The more modern post-harvest processing that can be accomplished as a result of using technology in agriculture is yet another advantage of using technology in agriculture. Handling harvested products after they have been harvested using the appropriate technology can improve both the quality and the durability of the products harvested. This is because the product has a longer life span, the process of decomposition can be slowed down, and the overall appearance of the product improves, making it more appealing to consumers. The process of distributing agricultural products can also be improved to a greater extent by making better use of technology in agriculture. This is due to the development of harvesting methods that are progressively more efficient and quicker, which allows for the crops to be immediately distributed to consumers.

3.8 NEGATIVE IMPACTS OF AGRICULTURAL TECHNOLOGY

The application of technology in agricultural settings comes with a number of additional benefits. Even with its many benefits, the use of technology has not been without its share of drawbacks. For instance, natural fruit is a very good source of vitamins and nutrients for the body, and it's also delicious. However, in today's world, due to the ever-evolving nature of agricultural technology, many fruits have been altered from their natural state through the processing and packaging methods that they undergo. Our ancestors used to plant plants by providing them with fresh water and compost on a daily basis. These days, however, because the soil conditions are not the same as they were in the past, we need to use a pump to provide fresh water to the plants. Pests of plants are becoming more varied over time; as a result, it is essential to make use of pesticides in order to both deter and eliminate pests of plants. The use of pesticides is evidence of the progression of technology; however, if you eat fruit that has been contaminated with pesticides and then go on to consume more fruit, you will almost certainly put your health in jeopardy. Pests that have not been eradicated entirely will become more dangerous if pesticides continue to be used. The relatively high cost is yet another effect of utilizing technological tools. Because of the high costs, the crops will undoubtedly have a high value when they are sold. The underprivileged will be affected as a result of this, especially when crops that are relatively expensive are essential to human survival. It will be challenging for those who are less fortunate to satisfy their fundamental requirements.

The use of technology also has an adverse effect on the surrounding natural environment, among other things. Several different technologies that continue to get their energy from fuel oil have consequences for the environment in the form of exhaust emissions. This will have an impact on the surrounding environment in the area of agricultural land if it continues for an extended time. In addition, the impact of the application of technology has an effect on the quality of the soil that is used for agricultural purposes. The quality of the soil on agricultural land can suffer if modern heavy equipment technology is utilized to an excessive degree and for an extended time. As a consequence of this, it is necessary to suspend the use of land for a specific amount of time in order to bring the soil quality back up to an acceptable level. Another thing that is related to the underground water structure is a negative impact that can be attributed to the use of technology. Withdrawing an excessive amount of water in an improper manner can have a disastrous effect on the structure of water found in the soil (Stoyanova and Harizanova, 2019).

3.9 CONCLUSION

The application of human knowledge for the purpose of improving living conditions can be defined as technology. Technology is inseparable from human resources and natural resources when it comes to the process of building a nation's independence; however, this process can only be completed if people have a firm grasp on technology. Therefore, agricultural technology is an effort made by humans to utilize science for the benefit and welfare of the agricultural industry. In addition, this type of agricultural technology comes in a variety of forms, the most common of which are traditional agricultural technology and modern agricultural technology.

The term "modern agricultural technology" refers to a method of farming that is carried out with the assistance of implements that incorporate contemporary technological developments. When working with farmers who use tractors to plow fields; for instance, the modern method of using tractors is more efficient and can shorten the time it takes to plow fields when compared to the traditional method of using buffalo or hoeing. The use of modern tools can reduce the amount of time it takes to plow fields as well as increase the amount of efficiency it provides. In general, the most prominent examples of modern technology are those that pertain to the application of seeds, fertilizers, pest control, and other similar practices.

There is not just one type of agricultural technology; there are both modern and traditional agricultural technologies. The term "traditional agricultural technology" refers to a method of farming that makes use of basic implements and is practiced on farms even today. When it comes to traditional agricultural technology, the level of efficiency is still not as high as it could be. This means that in order to get the best possible results from using traditional tools, a significant amount of effort and energy must be invested. Tillage with hoes is one example of the use of traditional technological equipment. Another is the process of plowing rice fields with the power of buffalo. Other examples include doing manual irrigation from streams, planting with human power, and fertilizing with manual sprays.

REFERENCES

Adhitya, Nanang Ika. (2018). Prototype Pest Repellent Bird in the Ricefield Based Arduino Uno. *Electronic Journal of Electronic Engineering Education*, 7(3), 67–78.

Aflitto, Nicholas, and DeGomes, Tom. (2020). *Sonic Pest Repellents*. Arizona: University of Arizona.

Alhassan, Yohanna J., Haruna, Yusuf, Muhammad, Muhammad A., SK, Fiddausi. (2019). Economics of Bio-Based Fertilizer in Improving Crop Productivity Through Extension Services Delivery. *International Journal of Agriculture and Plant Science*, 1(4), 10–13.

Hans, V. Basil. (2018). Water Management in Agriculture: Issues and Strategies in India. *International Journal of Development and Sustainability*, 7(2), 578–588.

Kumar, Khausal, Shukla, U.N., Kumar, Dharmendra, Pant, Anil Kumar, and Prasad, S.K. (2013). Bio-Fertilizers for Organic Agriculture. *Popularkheti*, 1(4), 91–95.

Mahdi, S. Sheraz, Hassan, G.I., Samoon, S.A., Rather, H.A., Dar, Showkat A., and Zehra, B. (2010). Bio-Fertilizers in Organic Agriculture. *Journal of Phytology*, 2(10), 42–54.

Manida, M., and Ganeshan, M.K. (2021). New Agriculture Technology in Modern Farming. *International Journal of Management Research and Social Science*, 8(1), 109–114.

Nair, Pratap, Nithiyananthan, K., and Dhinakar, P. (2017). Design and Development of Variable Frequency Ultrasonic Pest Repeller. *Journal of Advanced Research in Dynamical and Control Systems*, 9(12), 22–34.

Onah, C.I., and Iloka, C.M. (2013). Construction of an Ultrasonic Pest Repeller. *Journal of Space Science & Technology*, 2(1), 1–15.

Rehman, Abdul, Jingdong, Luan, Khatoon, Rafia, and Hussain, Imran. (2016). Modern Agricultural Technology Adoption its Importance, Role and Usage for the Improvement of Agriculture. *American-Eurasian Journal of Agricultural & Environmental Sciences*, 16(2), 284–288.

Sadiku, Matthew N.O., Ashaolu, Tolulope J., and Musa, Sarhan M. (2021). Emerging Technologies in Agriculture. *International Journal of Scientific Advances*, 1(1), 31–34.

Savei, Serpil. (2012). An Agricultural Pollutant: Chemical Fertilizer. *International Journal of Environmental Science and Development*, 3(1), 77–80.

Stoyanova, Zornitsa, and Harizanova, Hristina. (2019). Impact of Agriculture on Water Pollution. *AGROFOR International Journal*, 4(1), 111–118.

Syahminan. (2017). Prototype of Bird Repellent on Rice Plants Based on Arduino Microcontroller. *SPIRIT Journal*, 9(2), 26–34.

Veronica, N., Guru, Tulasi, Thatikunta, Ramesh, and Reddy, Narender. (2015). Role of Nano Fertilizers in Agricultural Farming. *International Journal of Environmental Science and Technology*, 1(1), 1–3.

Yaseen, Muhammad, Karim, Mujahid, Luqman, Muhammad, Mahmood, Muhammad Umer. (2019). Role of Agricultural Journalism in Diffusion of Farming Technologies. *Journal of Agricultural Research*, 57(4), 289–294.

Zarate, E., Aldaya, M., Chico, D., Pahlow, M., Flachsbarth, I., Franco, G., Zhang, G., Garrido, A., Kuroiwa, J., Pascale-Palhares, J.C., and Arévalo, D. (2014). *Water for Food and Wellbeing in Latin America and the Caribbean*. London: Routledge.

4 Off-Axis Reflectors for Concentrated Solar Photovoltaic Applications

Ramsundar Sivasubramanian,
Chockalingam Aravind Vaithilingam,
Sridhar Sripadmanabhan Indira, Chong Kok
Keong, and Samsul Ariffin Abdul Karim

CONTENTS

4.1 INTRODUCTION

Solar energy has long been exploited as a cheap, effectively omnipresent source of renewable energy. It was primarily exploited as a source of heat till the 19th century when physicist Edmond Becquerel discovered the photovoltaic effect in 1839. In the present day, solar energy has firmly been established among the leading renewable sources of heat and electricity. Accordingly, the field of solar power is broadly classified based on the mode of energy harnessed—solar

DOI: 10.1201/9781003367819-4

photovoltaics, solar thermal, and any combination of the two in varying proportions thereof (Muhammad-Sukki et al., 2010; Sripadmanabhan Indira et al., 2020).

Solar photovoltaic devices are based on the photovoltaic effect—energy from photons from solar radiation incident on an appropriate semiconductor material enables electrons in the material's valence band to bridge the gap to the conduction band and, hence, drive current through a circuit. Depending on the way sunlight is incident on solar panels, PV systems are further classified into flat-plate PV and concentrated PV. In flat-plate PV, sunlight hits the panel directly with no manner of concentration involved; that is, the intensity of light on the panel is nearly equal to the Global Horizontal Irradiance (GHI) at the location. In contrast, CPV systems utilize an assemblage of components, such as lenses and/or mirrors, to concentrate irradiation falling on a larger area onto a smaller receiver. The degree of concentration can have the following range: 2–100 (LCPV), 100–300 (MCPV), to 300–1,000+ (HPCV) (Kalogirou, 2014). This dichotomy results in distinct characteristics for each of the two types of systems. Flat-plate PV utilizing single-junction solar cells offer limited efficiencies of up to 20%, while CPV systems using multi-junction cells (Guter et al., 2009) can go as high as 30–40% (Green et al., 2020; Osterwald & Siefer, 2016). However, the concentration ratios involved in CPV systems necessitate the use of solar tracking while it's not mandatory for flat-plate PV (Emodi & Boo, 2015; Ogunleye, 2011). When it comes to cost, CPV systems potentially hold the advantage since they can substitute cheaper mirrors/lenses installed over a given area to achieve an equivalent power output compared to flat-plate PV systems needing to cover a larger area with expensive solar cells, resulting in semiconductor material savings (Abdalla, 2003; Emam et al., 2016). The concentrators used in solar thermal systems are usually fabricated with mirrors, while CPV concentrators employ clear plastics or glass. In certain locations with good solar exposure—that is, high Direct Normal Irradiance (DNI)—these cost savings can enable the installation of more expensive, high-efficiency cells (Kamath et al., 2019), which in turn can potentially reduce the levelized cost of electricity (LCOE) of CPV systems to levels competitive to standard flat-PV technology (Ejaz et al., 2021; Philipps et al., 2015).

In this chapter, the principles behind off-axis concentrators are discussed, along with the design and analysis of a small-scale off-axis concentrator for CPV application in an urban setting.

4.2 PARABOLIC CONCENTRATORS

There exist several commonly used reflector geometries for concentrator design in CPV applications. They range from Fresnel lenses, v-troughs, to parabolic reflectors (Antón et al., 2003; Kussul et al., 2008; Zanganeh et al., 2012). Parabolic reflectors consist of contours shaped based on a parabola. Paraboloidal dishes, such as solar cookers, have been widely used to concentrate solar energy and convert it into medium high-temperature heat (Eck & Hennecke, 2009). Solar parabolic dish concentrators may produce high temperatures for solar cooking, solar water heating (Badran et al., 2010), solar thermal energy, and solar thermal steam production, among other solar thermal uses (Cui et al., 2003; Kodama, 2003). The focal point of a parabolic dish concentrator is a parabolic reflecting surface (dish) with a solar thermal receiver. Parabolic dish concentrators can produce far higher temperatures than other types of solar concentrators, making them suitable for high-temperature processes (Lovegrove et al., 2011). The receiver of a parabolic dish concentrator must be well-engineered in order to achieve high temperatures with little heat loss. Many theoretical and practical studies on parabolic dish concentrators for small to big industrial applications have been done in recent years (Mawire & Taole, 2014). The fundamental property of a parabola dictates that all radiation incident parallel to the parabolic axis tends to be focused at the focal point of a parabola. This is the principle behind the operation of a parabolic reflector.

Parabolic concentrators are further classified into imaging and non-imaging concentrators. Imaging concentrators consist of linear parabolic trough collectors (linear concentrators) and parabolic dishes (point concentrators), while compound parabolic reflectors constitute the non-imaging type. Both types are distinguished by the concentrated image formed at the receiver—the former results in a highly focused and well-defined image, while the latter generates a relatively more diffused image. Also, imaging concentrators function only with tracking, while non-imaging concentrators offer a certain degree of flexibility by still concentrating incident radiation falling within a certain acceptance angle with respect to the reflector. But imaging concentrators generate higher concentration ratios compared to non-imaging concentrators (Ghani, 2014).

4.2.1 OFF-AXIS PARABOLIC CONCENTRATORS

This chapter focuses on a particular subset of parabolic concentrators—namely, the off-axis parabolic reflectors (Figure 4.1). Off-axis reflectors consist of paraboloid profiles sectioned off from a larger master paraboloid such that the sectioned surface is offset from the axial plane of the generating paraboloid. In effect, this operation produces a parabolic reflector surface, which is separated from the parabolic axis when viewed from the top plane from where radiation is incident on the system. This enables the receiver in off-axis systems to be placed away from the reflector surface causing minimal/no overlap between the two entities along the plane of incidence. Typically, the receiver assembly of a parabolic point concentrator is situated on a plane parallel to the reflector plane, causing the reflector to suffer from a loss known as shading loss due to the overlap of the two surfaces. This loss can be eliminated using off-axis reflectors.

An off-axis reflector has a higher surface area for a given projected normal area than a standard parabolic point concentrator. Furthermore, the parent paraboloid's equation, which also defines the aperture width and rim angles, affects the relative positioning of the focus point on the off-axis reflector. Depending on the sectioning plane used for generating the off-axis section, its curvature varies. The closer the sectioning plane is to the vertex of the parent paraboloid, the

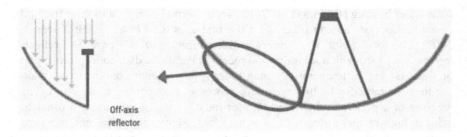

FIGURE 4.1   A schematic showing the difference between a parabolic dish concentrator and an off-axis concentrator.

greater the curvature of the off-axis section. Similarly, as the sectioning plane is pushed further away from the vertex, the curvature of the off-axis section is reduced, with minimal curvatures observed at the sections closest to the rim of the parent paraboloid. This characteristic results in another advantage of off-axis reflectors compared to conventional parabolic point concentrators—their reduced curvatures improve the ease of fabrication, and the resultant mirrors tend to be less prone to distortions arising from variable stresses in curved sections during manufacturing. But there is a trade-off to be made for the reduced curvatures—the lower the curvature, the farther the distance between the reflector and the receiver. Thus, the design of an off-axis reflector is constrained primarily by the factors of the concentration ratio required, permissible curvature range, and spacing between the receiver and reflector.

4.2.2   SOLAR TRACKING FOR IMAGING CONCENTRATORS

In solar tracking, the whole movement of this system is accomplished by modifying the plane's location in relation to the sun's direction. As a result, sunlight will be focused on the solar concentrator's reception area. With a process of trial and error, adjusting the size of the reflector and manually shifting its direction are also possible but will take a long time. To resolve all these flaws, dual-axis tracking systems were designed and improved to the point where they can now be found in modern solar concentrator systems. The tracking system uses two axes since the receiver base may move up and down as well as left and right, making it very easy to receive sunlight from either side (Agarwal, 1992). The tracking system is designed to track the sun in order to capture as much solar radiation as possible in terms of thermal energy. There are two types of tracking: single-axis tracking and dual-axis tracking. Single-axis solar trackers usually follow the sun's movement from east to west or north to south on a daily basis. Dual-axis solar trackers, in contrast, track both the altitude and azimuthal angles of the sun; that is, the sun's movements from east to west as well as north to south. Furthermore, these systems may be classified into two categories, depending on their operational principles: manual tracking and automated or sensor-based tracking. It can also be divided into active and passive types of trackers. Furthermore, it may be classified as a single-axis or dual-axis tracker, depending on the rotational axis (Sahu et al., 2021).

4.2.3 Multi-Junction Solar Cells

Another integral component in CPV systems is the multi-junction solar cell. These cells, which can operate at higher irradiance and temperature compared to conventional single-junction cells, enable the operation of CPV systems. The high-efficiency, high-concentration CPV (HCPV) technology has aroused attention as concentrator cells have attained ever-increasing efficiencies. Several multijunction cell designs have currently demonstrated efficiencies in the 40% range. The particular structures inside each of these designs might be differentiated further. Solar junction now holds the efficiency record at 43.5% (Espinet González et al., 2012). This cell's lowest junction is constructed from a dilute nitride alloy that was grown using molecular beam epitaxy. NREL and Emcore have produced inverted metamorphic, 3-junction cells with efficiencies of 42.6% and 42.4%, respectively (Geisz et al., 2012). Spire achieved 42.3% efficiency using a bi-facial technique (GaInP/GaAs on the front and GaInAs on the back of a GaAs wafer) (Wojtczuk et al., 2011).

Because the expensive solar cells are replaced with less expensive structural steel supporting mirrors or lenses, CPV systems are anticipated to be relatively low-cost power producers. Early CPV systems, however, increased by 15% to 20%. The efficiency of today's CPV systems, which use extremely efficient III-V silicon solar cells, is reaching 29%. Installed CPV systems are now equivalent to utility-scale flat-plate PV systems in terms of cost. However, because these efficiencies are considerably below the physical constraints for converting sunlight into energy, further development of III-V solar cells, which are presently roughly 42% efficient is still achievable (Robert McConnell & Vasilis Fthenakis, 2012). The GaInP and GaInAs sub-cells can be grown on a metamorphic buffer in 3-junction GaInP/GaInAs/Ge metamorphic (MM) solar cells so that they are lattice-matched (LM) to each other but lattice-mismatched to the Ge growth substrate and sub-cell (King et al., 2007). The upper two junctions of the MM cell may be constructed with narrower bandgaps than the LM case due to the mismatched structure, resulting in a superior bandgap combination for converting the solar spectrum (Cotal et al., 2009).

There are several methods for lowering the levelized costs of power generated by photovoltaics. On the one hand, module costs fall as a result of economies of scale, reduced material and energy usage, and the utilization of low-cost materials. Increased module efficiency, on the other hand, can lower system costs while also allowing for smaller systems and more efficient use of space. As a result, all components of an HCPV system must be further improved in order to achieve 12 maximum efficiencies. Large strides have been made, particularly in the field of III-V multi-junction solar cells, where separate groups achieved record efficiencies of over 41% in 2009 (Guter et al., 2009; King et al., 2007). Despite the high concentration levels, the solar cell can account for up to 20% of the total cost of an HCPV system (McConnell et al., 2005). As a result, a highly efficient multi-junction solar cell is a critical component for future energy cost reduction. Multiple stacking of solar cells with increasing bandgap energies improves total device efficiency because the solar spectrum is more productively used. At the UPM Madrid, utilizing Ga0:51In0:49P and GaAs sub-cells, a record value of 32.6% under 1,000 suns (AM1.5d) was attained for a monolithic III-V dual-junction solar cell (García et al., 2009).

As a result of the predicted rise in cell efficiency, research activities for the creation of III-V multi-junction solar cells with more than three sub-cells have increased dramatically in recent years (*Conference Record of the 2006 IEEE 4th World Conference on Photovoltaic Energy Conversion : May 7–12, 2006, Waikoloa, Hawaii; Incorporating 32nd IEEE Photovoltaic (PV) Specialist Conference and 16th Asia/Pacific (International) Photovoltaic (PV) Science and Engineering Conference, IEEE World Conference on Photovoltaic Energy Conversion; 4*, 2006; Zhu et al., 2016). It is a well-known fact that the optimal efficiency of a multi-junction solar cell grows with the number of p n-junctions in a given spectrum (France et al., 2016). Multi-junction solar cells, however, are known to be extremely sensitive to variations in the sun spectrum (Faine et al., 1991; Meusel et al., 2002). The researchers Meusel et al. (2002) investigated several solar energy strategies for generating power. This study looked at both direct and indirect electricity-generating methods. PV modules were utilized for direct solar energy conversion, while various optical devices were employed for indirect solar energy harvesting. The advantages and disadvantages of various solar collector technologies, such as parabolas, trough collectors, Fresnel lenses, and central towers, were briefly examined. PV technologies outperform concentrated solar power plants like CSP plants in small-scale electricity generation, according to studies examined. CSP systems, in contrast, are superior in terms of economic return. PV and CSP systems have been discovered to have maintenance costs of 2% and 1%, respectively. When solar concentrators are compared, the parabolic concentrator is shown to be more effective than the Fresnel lens concentrator. Furthermore, parabolic dish concentrators were shown to be more successful when solar energy conversion was required at very high temperatures.

Many countries, including Malaysia, are now putting a greater emphasis on green technologies and renewable energy. Renewable energy, such as solar power generators, might generate the majority of the world's electricity within 50 years, according to a 2011 prediction by the International Energy Agency, drastically lowering greenhouse gas emissions that hurt the environment. Solar energy is by far the most plentiful of all renewable energy sources (Faine et al., 1991). Malaysia receives a lot of sunlight and solar radiation—roughly six hours of sunlight every day on average. Malaysia has an annual average of 4.21–5.56 kWh/m^2 of daily solar irradiation and more than 2,200 hours of sunlight each year (Affandi et al., n.d.; Ahmadi et al., 2018; Muzathik et al., 2010). As a result, Malaysia can pursue solar energy aggressively to meet some of its energy demands and, consequently, improve energy security.

4.3 METHODOLOGY

This section describes the methodology employed in this work for the design and development of the small-scale off-axis CPV system.

4.3.1 OFF-AXIS REFLECTOR GEOMETRY MODELING

The geometry of the off-axis reflector was modeled using CAD software (SOLIDWORKS). An equation-driven curve was used to generate the profile curvature of the parent paraboloid. The parabolic equation used was of the following form:

$$y^2 = 4.f.x \qquad (4.1)$$

where y and x are the coordinates along two perpendicular axes and f is the focal length of the parabola (i.e., the distance between the vertex and focal points).

The curve generated was then revolved about its axis and thickened to generate the parent paraboloid. The off-axis section was sectioned from the parent paraboloid by defining a cutting plane parallel to the incident plane. A circular profile was defined on the cutting plane at a fixed distance from the point where the parabolic axis passed through the plane. Then an extruded cut was used to section the off-axis reflector. The receiver profile for the parabola was sketched at the focal plane and extruded to get the receiver. The final geometries were exported as STEP files for further processing. Figure 4.2 and Figure 4.3 show two different views of the system geometry. The elliptical off-axis reflector used in this study had a major axis diameter of 650 mm, with the off-axis section inclined at 45° with respect to the parabolic axis. A conventional parabolic reflector geometry with the receiver surface overlapping the reflector was also designed for a comparative analysis with the off-axis system.

To obtain the most concentrated solar radiation, a receiver must be placed at the focal point of the parabola. The focal point is perpendicular to the reflector surface. It is the diameter of the parent paraboloid dish that is relevant, not the off-axis reflector characteristics, in the determination of the focal length and image size. The minimum receiver diameter (D) required to capture the entire image from a parabolic reflector is given by:

$$D = 2r_r \sin(\theta_m) \qquad (4.2)$$

where, r_r is the maximum mirror radius and θ_m is the half acceptance angle for the reflector.

Also, the focal distance (f) is related to the mirror radius (r) and the angle between the focus of a reflected beam and the collector axis (φ) as follows:

$$r = 2f/(1 + \cos(\varphi)) \qquad (4.3)$$

FIGURE 4.2 Isometric view of the off-axis reflector geometry—this image shows the reflector and receiver components of the off-axis reflector–based CPV system.

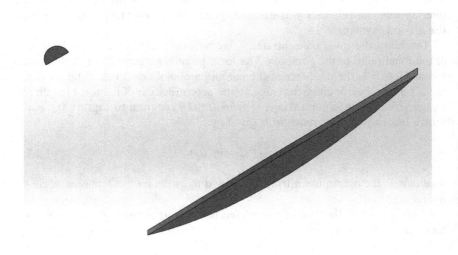

FIGURE 4.3 Side view of the model geometry for the off-axis reflector–based CPV system.

The image size on the receiver plane increases from $2f\sin(\theta_m)$ to $2r_r\sin(\theta_m)/\cos(\varphi_r + \theta_m)$ when r increases from f to r_r and φ changes from 0 to φ_r. Another seminal relation that's indispensable in the design of parabolic reflectors correlates the aperture width (W_a) to the focal length and the rim angle (φ_r). It is given by the following equation:

$$W_a = 4f.\tan(\varphi_r/2) \tag{4.4}$$

4.3.2 Monte Carlo Ray-Tracing Simulation

The optical performance of the off-axis reflector was simulated using Monte Carlo ray tracing (Bendt & Rabl, 1981; Daly, 1979) in the TracePro software (Figure 4.4 and Figure 4.5). Monte Carlo ray tracing involves defining a source,

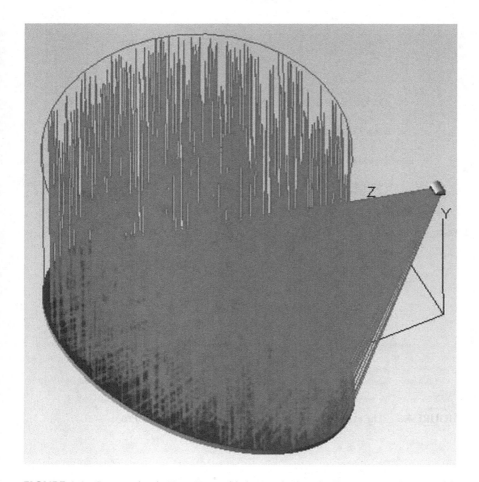

FIGURE 4.4 Ray tracing in TracePro—this image depicts the TracePro workspace with a preview render of the results of a ray trace.

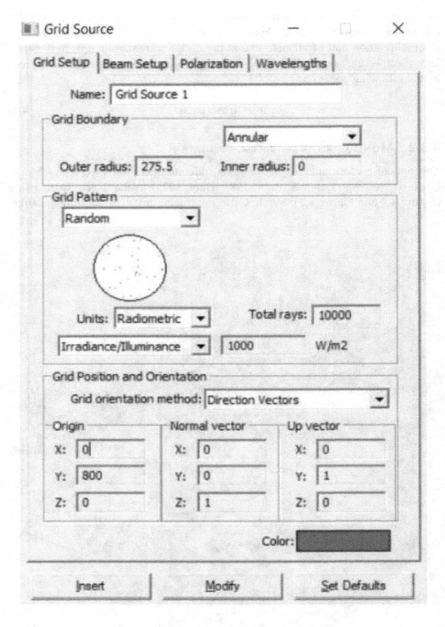

FIGURE 4.5   The radiation (grid) source setup window in TracePro.

TABLE 4.1

The Irradiation Source Parameters in TracePro

Parameter	Values
Grid Boundary	
Outer radius	275.5 mm
Inner radius	0 mm
Grid Pattern	
Units	Radiometric
Total rays	100,000
Irradiance/Illuminance	1,000 W/m²
Grid Position and Orientation	
Origin	$X = 0, Y = 800, Z = 0$
Normal vector	$X = 0, Y = 0, Z = 1$
Up vector	$X = 0, Y = 1, Z = 0$

specifying the optical properties of the model geometry, and running the ray trace. The ray-tracing process involves the generation of randomly populated rays from the defined source plane whose varying interactions (reflection, transmission, absorption, refraction, scattering, etc.) with objects in the model space depend on an event probability matrix shaped by the predefined optical properties. Then one of the surfaces/bodies, usually the receiver, in the model is selected, and the results of its interaction with incident radiation are analyzed in the form of 2D/3D irradiation plots.

There are three types of irradiation sources; namely, the surface source, grid source, and file source, available in TracePro to simulate varying sources (Table 4.1). For this work, to approximate solar irradiation, an annular grid source with a radius of 275.5 mm was used. This radius corresponds to the dimension of the projected aperture for the off-axis reflector on the incidence plane. The source was overlaid at the calculated system aperture, exactly at the projection of the reflector surface on the incident plane. The source type was set as a radiometric source with an irradiance value of 1,000 W/m². The rays were set to diverge at an angle of 0.27° to replicate the mean solar divergence angle.

The total number of rays to be used for the analysis was determined by means of a ray independence test. The ray-tracing analysis for the model geometry was run multiple times at varying ray numbers. The resulting irradiation patterns on the receiver were analyzed, and the ray number was selected as the minimum number of rays at which the irradiation pattern results were stabilized. This demonstrated that the results of the ray tracing were unaffected by the number of rays generated from the source. For the current study, 10^5 rays yielded the best results. This test is essential to achieve the best results at the least computational cost.

The material properties of the objects in the model geometry were set as follows:

- Reflectivity of the off-axis mirror surface was set as 0.85 to simulate the reflectivity of the acrylic-based mirrors used in the study.
- The receiver was assumed as a perfect absorber (i.e., an absorptivity value of 1.0). This was done to ensure that all the irradiation falling on the receiver from the concentrator could be analyzed.

4.3.3 FABRICATION OF THE LAB-SCALE PROTOTYPE

In order to study the actual performance of the proposed off-axis system and compare it with the simulated performance, a lab-scale prototype was fabricated using a satellite television dish antenna as the base (Figure 4.6). A satellite TV dish also

FIGURE 4.6 Photograph of the reflector used in the lab-scale prototype—a satellite television dish converted into a solar concentrator by layering it with acrylic mirrors.

possesses the off-axis geometry needed for this study. A coating of flexible acrylic-based mirrors was applied to the dish to convert it into a parabolic mirror concentrator suitable for use with solar radiation. The receiver assembly consisting of an assemblage of multijunction cells covered by compound parabolic lenses for uniform irradiation dispersion and the heat sink was attached to the receiver holder in place of the satellite TV receiver. The lenses with compound parabolic profiles help distribute the concentrated irradiation uniformly over the surface of the multijunction cell, hence, avoiding hotspots and the associated cell damage and efficiency loss. The heat sink made of aluminum fins and copper channels helped limit the temperature of the solar cells during operation under high-intensity concentrated solar irradiation. This was accomplished by means of improved convective heat loss from the increased surface area (boundary layer) of the aluminum coupled with the effective heat transfer (conduction) along the copper channels and contacts from the multi-junction cells to the heat-spreading fins.

To enable receiver offset from the focal point, a combination of a slotted bar and hinge was used to mount the receiver assembly. The hinge allowed for variation in the receiver angle, while the slotted bar and fastener mechanism allowed vertical adjustment of the receiver assembly (Figure 4.7 and Figure 4.8).

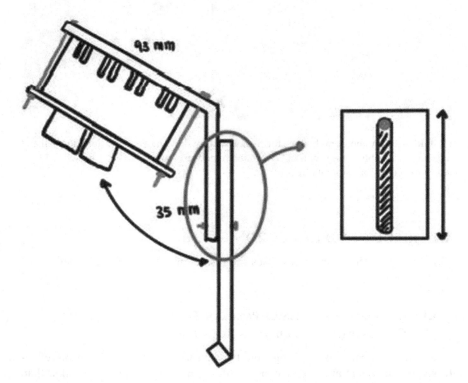

FIGURE 4.7 A schematic of the adjustable receiver mount used in the lab-scale prototype.

FIGURE 4.8 A photograph of the receiver assembly showing the multi-junction cell and heat sink assembly, along with the adjustable receiver mount. Each cell is covered by a lens with a compound parabolic profile.

4.4 RESULTS AND DISCUSSION

This section of the chapter discusses the results of the studies described in the sections beforehand.

4.4.1 COMPARATIVE ANALYSIS OF PARABOLIC DISH CONCENTRATOR AND OFF-AXIS REFLECTOR

The irradiation patterns resulting from two similar ray-tracing studies on reflectors with equivalent areas were considered. All factors for both cases other than the reflector geometry were identical. The irradiation patterns in Figure 4.9 and Figure 4.10 show the results of the ray-tracing simulation for parabolic point

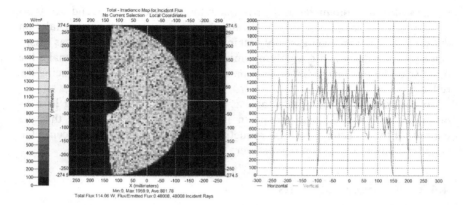

FIGURE 4.9 Irradiation map for a conventional parabolic dish concentrator—the map represents the right half of the reflector; note the circular shadow along the left (center of the reflector) periphery of the map.

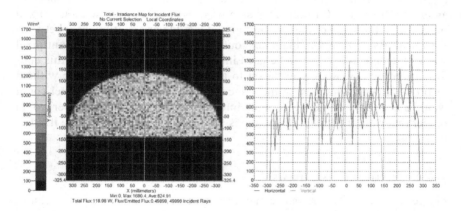

FIGURE 4.10 Irradiation map for the proposed off-axis reflector–based CPV system—the image shows the irradiation pattern on half of the off-axis reflector; note the uniform irradiation all over the reflector surface.

concentrators and off-axis reflectors, respectively. In the case of conventional parabolic dishes, there exists a dark spot at the place where the receiver assembly overlaps with the reflector when viewed along the incidence plane. This loss occurs as a result of the shadow of the receiver falling on the reflector and, hence, rendering any reflector area covered by the shadow moot. This loss in the reflector area can result in up to 42 W/m² of irradiance loss, depending upon the receiver assembly geometry and location. That is enough power to run small-scale gadgets, such as gaming consoles and personal computing devices. In addition to

the loss of power, the material used to make the reflector is also rendered inef-
fective in the areas affected by this shading loss. Any irradiation that falls on
the surface of the receiver assembly facing the sun mostly goes unused as losses
since receiver designs are highly optimized and directional, and can make the
best use of the incident radiation only on a designated receiver surface facing the
reflector(s).

In contrast, an off-axis reflector was able to eliminate the shading loss, as
observed from the irradiation pattern in Figure 4.10. This ensured that the
entirety of the reflector area was productive and contributed to the concentrated
solar irradiation focused on the receiver.

4.4.2 RAY INDEPENDENCE TEST

The average irradiance values obtained from repeating the same ray-tracing
analysis with varying ray numbers are listed in Table 4.2. The result of the ray
independence test showed that 10^5 rays was the optimum ray count for this
study. Prior to running the ray independence test, the pixel count for the irradiation plot was fixed so that each pixel on the irradiance plot corresponded to
an area of 0.1 mm × 0.1 mm. This step was essential because without a fixed
pixel count, the scale of the irradiation plot would vary continuously between
each trial. By convention, the pixel count is set to match the least count/pixel
size of the detector/sensor to be used in the practical trials. Barring this condition, the pixel count could be set to a reasonable minimum value without
disproportionately increasing the ray count required. Too high a pixel count
would make the ray-tracing resource intensive, while too low a number would
render the resultant data grainy and possibly result in detail loss in the irradiation plots.

4.4.3 RECEIVER OFFSET TRIALS

Figure 4.11 shows the irradiance map for a receiver placed at the focal point of
the proposed off-axis system. It shows a typical elliptical concentrated image
characteristic of off-axis parabolic reflectors. For this case, the value of peak

---

TABLE 4.2

Ray Independence Test

Ray Count	Peak Irradiance (W/m²)	Average Irradiance (W/m²)
1,000	7.0138 e+6	3.7347 e+5
10,000	6.8234 e+6	3.7004 e+5
100,000	6.4777 e+6	3.6919 e+5
1,000,000	6.4778 e+6	3.6919 e+5

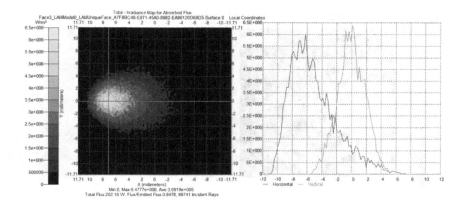

FIGURE 4.11 Irradiance map for the off-axis reflector–based CPV system with the receiver placed at the focal point.

irradiance and the average irradiance were observed to be 6.4777 e+6 W/m^2 and 3.6919 e+5 W/m^2, respectively. These values were quite high and unsuitable for use with a small-scale CPV system. In addition, the concentrated image obtained was highly non-uniform throughout the cross-section, with a glaring hot spot at the center of the elliptical image. Such hot spots are sought to be avoided in CPV design to ensure the long-term viability of the multi-junction cells used in the receiver. The local intensity profiles of the irradiation map also portray the same—a prominent peak was observed along both the vertical and horizontal axes passing through the center of the image.

To mitigate these issues of high-concentration intensity and non-uniform concentrated image intensities, offsetting the receiver from the focal plane was proposed. This would result in the converging radiation from the reflector being intercepted along a plane closer to the reflector surface, hence, producing a larger and more scattered image. The results of offsetting the receiver to a parallel plane of 25 mm and 50 mm are shown in Figure 4.12 and Figure 4.13, respectively.

It can be observed that the image size increased progressively with increasing offset distance. Also, the concentrated image intensities over the image cross-section improved with rising offset distance, too, as evidenced by the irradiance intensity contours accompanying the irradiation plots. Table 4.3 summarizes the results of these studies with the peak and average irradiance values in each case—the overall values were observed to be lesser than in the case of placing the receiver at the focal point.

Thus, offsetting the receiver was demonstrated to be a viable strategy in the design of off-axis reflector-based CPV systems, where the offset distance can also be considered one of the variables to be tweaked to attain the necessary concentration ratio and image intensities.

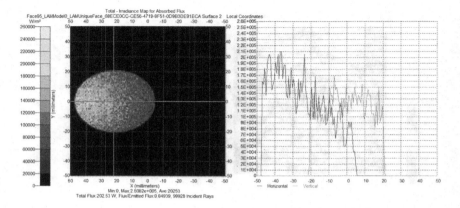

FIGURE 4.12 Irradiance map for the off-axis reflector–based CPV system with the receiver offset 25 mm from the focal plane.

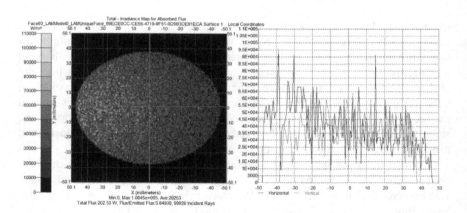

FIGURE 4.13 Irradiance map for the off-axis reflector–based CPV system with the receiver offset 50 mm from the focal plane.

TABLE 4.3
Irradiance Results for the Receiver Offset Trials

	Focal Plane	25 mm Offset	50 mm Offset
Max flux (W/m^2)	6.4777 e+6	2.5082 e+5	1.0945 e+6
Average flux (W/m^2)	3.6919 e+5	20,253	20,253
Total flux (W)	202.16	202.53	202.53
No. of incident rays	99,741	99,928	99,928

4.5 CONCLUSION AND SUMMARY

The results of this study demonstrated the performance and feasibility of employing off-axis reflectors for CPV systems. Using off-axis reflectors, the shading losses plaguing conventional parabolic dish concentrators can be alleviated or even eliminated in entirety. Also, off-axis reflectors enable the best use of available reflector surface area, and even reflectors with lower curvatures can be used improving ease of fabrication. The use of the method of receiver offset was also discussed as a means of improving the concentrated image intensity and uniformity.

While this study has demonstrated the advantages of off-axis reflector-based CPV systems, they are still a way off from becoming viable solutions ready to be implemented in practice. There is still potential for further investigation of this idea, such as optimizing the receiver placement, determining the most suitable concentration ratio, designing an appropriate tracking system, and further improving concentrated image uniformity. The last issue listed can potentially be solved by the use of an additional secondary reflector at the location of the primary reflector's receiver. This receiver would be a scaled-down version of the primary reflector and, if oriented appropriately, could convert the elliptical focused image from the primary reflector into a circular image with greater uniformity.

REFERENCE

Abdalla, F. K. (2003). *Cost per kWh Produced and Payback Time of a PV-Solar-Thermal-Combined System at Different Locations in New Zealand*, 9.

Affandi, R., Ghani, M. R. A., & Gan, C. K. (n.d.). A Review of Concentrating Solar Power (CSP) in Malaysian Environment. *International Journal of Engineering and Advanced Technology (IJEAT)*, 3(2), 5.

Agarwal, A. K. (1992). Two Axis Tracking System for Solar Concentrators. *Renewable Energy*, 2(2), 181–182. https://doi.org/10.1016/0960-1481(92)90104-B

Ahmadi, M. H., Ghazvini, M., Sadeghzadeh, M., Alhuyi Nazari, M., Kumar, R., Naeimi, A., & Ming, T. (2018). Solar Power Technology for Electricity Generation: A Critical Review. *Energy Science & Engineering*, 6(5), 340–361. https://doi.org/10.1002/ese3.239

Antón, I., Pachón, D., & Sala, G. (2003). Characterization of Optical Collectors for Concentration Photovoltaic Applications. *Progress in Photovoltaics: Research and Applications*, 11(6), 387–405. https://doi.org/10.1002/pip.502

Badran, A. A., Yousef, I. A., Joudeh, N. K., Hamad, R. A., Halawa, H., & Hassouneh, H. K. (2010). Portable Solar Cooker and Water Heater. *Global Conference on Renewables and Energy Efficiency for Desert Regions (GCREEDER 2009)*, 51(8), 1605–1609. https://doi.org/10.1016/j.enconman.2009.09.038

Bendt, P., & Rabl, A. (1981). Optical Analysis of Point Focus Parabolic Radiation Concentrators. *Applied Optics*, 20(4), 674–683. https://doi.org/10.1364/AO.20.000674

Conference Record of the 2006 IEEE 4th World Conference on Photovoltaic Energy Conversion: May 7–12, 2006, Waikoloa, Hawaii; Incorporating 32nd IEEE Photovoltaic (PV) Specialist Conference and 16th Asia/Pacific (International)

Photovoltaic (PV) Science and Engineering Conference, IEEE World Conference on Photovoltaic Energy Conversion; 4. (2006). IEEE Operations Center, www.tib. eu/de/suchen/id/TIBKAT%3A524855471

Cotal, H., Fetzer, C., Boisvert, J., Kinsey, G., King, R., Hebert, P., Yoon, H., & Karam, N. (2009). III—V Multijunction Solar Cells for Concentrating Photovoltaics. *Energy & Environmental Science*, 2(2), 174–192. https://doi.org/10.1039/B809257E

Cui, H., Yuan, X., & Hou, X. (2003). Thermal Performance Analysis for a Heat Receiver Using Multiple Phase Change Materials. *Applied Thermal Engineering*, 23(18), 2353–2361. https://doi.org/10.1016/S1359-4311(03)00210-2

Daly, J. C. (1979). Solar Concentrator Flux Distributions using Backward Ray Tracing. *Applied Optics*, 18(15), 2696–2699. https://doi.org/10.1364/AO.18.002696

Eck, M., & Hennecke, K. (2009). Heat Transfer Fluids for Future Parabolic Trough Solar Thermal Power Plants. In D. Y. Goswami & Y. Zhao (Eds.), *Proceedings of ISES World Congress 2007* (Vol. I—Vol. V) (pp. 1806–1812). Springer, Berlin, Heidelberg.

Ejaz, A., Babar, H., Ali, H. M., Jamil, F., Janjua, M. M., Fattah, I. M. R., Said, Z., & Li, C. (2021). Concentrated Photovoltaics as Light Harvesters: Outlook, Recent Progress, and Challenges. *Sustainable Energy Technologies and Assessments*, 46, 101199. https://doi.org/10.1016/j.seta.2021.101199

Emam, M., Ahmed, M., & Ookawara, S. (2016). Thermal Regulation Enhancement of Concentrated Photovoltaic Systems Using Phase-change Materials. 8th International Exergy, Energy and Environment Symposium (IEEES-8), May 1–4, 2016, Antalya, Turkey.

Emodi, N. V., & Boo, K.-J. (2015). Sustainable Energy Development in Nigeria: Overcoming Energy Poverty. *International Journal of Energy Economics and Policy*, 5(2), 580–597.

Espinet González, P., Barrigón Montañés, E., Ochoa Gómez, M., Barrutia Poncela, L., Orlando Carrillo, V., Algora del Valle, C., & Rey-Stolle Prado, I. (2012). Structural Challenges in Multijunction Solar Cells for Ultra-High CPV. 27th European Photovoltaic Solar Energy Conference and Exhibition, Frankfurt, Germany.

Faine, P., Kurtz, S. R., Riordan, C., & Olson, J. M. (1991). The Influence of Spectral Solar Irradiance Variations on the Performance of Selected Single-Junction and Multijunction Solar Cells. *Solar Cells*, 31(3), 259–278. https://doi.org/10.1016/0379-6787(91)90027-M

France, R. M., Dimroth, F., Grassman, T. J., & King, R. R. (2016). Metamorphic Epitaxy for Multijunction Solar Cells. *MRS Bulletin*, 41(3), 202–209. https://doi.org/10.1557/mrs.2016.25

García, I., Rey-Stolle, I., Galiana, B., & Algora, C. (2009). A 32.6% Efficient Lattice-Matched Dual-Junction Solar Cell Working at 1000 Suns. *Applied Physics Letters*, 94(5), 053509. https://doi.org/10.1063/1.3078817

Geisz, J. F., Duda, A., France, R. M., Friedman, D. J., Garcia, I., Olavarria, W., Olson, J. M., Steiner, M. A., Ward, J. S., & Young, M. (2012). Optimization of 3-Junction Inverted Metamorphic Solar Cells for High-Temperature and High-Concentration Operation. *AIP Conference Proceedings*, 1477(1), 44–48. https://doi.org/10.1063/1.4753830

Ghani, Mohd. R. A. (2014). The Influence of Concentrator Size, Reflective Material and Solar Irradiance on the Parabolic Dish Heat Transfer. *Indian Journal of Science and Technology*, 7(9), 1454–1460. https://doi.org/10.17485/ijst/2014/v7i9.31

Green, M. A., Dunlop, E. D., Hohl-Ebinger, J., Yoshita, M., Kopidakis, N., & Hao, X. (2020). Solar Cell Efficiency Tables (Version 56). *Progress in Photovoltaics: Research and Applications*, 28(7), 629–638. https://doi.org/10.1002/pip.3303

Guter, W., Schöne, J., Philipps, S. P., Steiner, M., Siefer, G., Wekkeli, A., Welser, E., Oliva, E., Bett, A. W., & Dimroth, F. (2009). Current-Matched Triple-Junction Solar Cell Reaching 41.1% Conversion Efficiency Under Concentrated Sunlight. *Applied Physics Letters*, 94(22), 223504. https://doi.org/10.1063/1.3148341

Kalogirou, S. (2014). *Solar Energy Engineering: Processes and Systems* (Second edition). Elsevier, AP, Academic Press is an Imprint of Elsevier. https://www.elsevier.com/books/solar-energy-engineering/kalogirou/978-0-12-397270-5

Kamath, H. G., Ekins-Daukes, N. J., Araki, K., & Ramasesha, S. K. (2019). The Potential for Concentrator Photovoltaics: A Feasibility Study in India. *Progress in Photovoltaics: Research and Applications*, 27(4), 316–327. https://doi.org/10.1002/pip.3099

King, R. R., Law, D. C., Edmondson, K. M., Fetzer, C. M., Kinsey, G. S., Yoon, H., Sherif, R. A., & Karam, N. H. (2007). 40% Efficient Metamorphic GaInP/GaInAs/Ge Multijunction Solar Cells. *Applied Physics Letters*, 90(18), 183516. https://doi.org/10.1063/1.2734507

Kodama, T. (2003). High-Temperature Solar Chemistry for Converting Solar Heat to Chemical Fuels. *Progress in Energy and Combustion Science*, 29(6), 567–597. https://doi.org/10.1016/S0360-1285(03)00059-5

Kussul, E., Baidyk, T., Makeyev, O., Lara-Rosano, F., Saniger, J., Bruce, N., & Mexico, D. (2008). Flat Facet Parabolic Solar Concentrator with Support Cell for One and More Mirrors. *WSEAS Transactions on Power Systems*, 3.

Lovegrove, K., Burgess, G., & Pye, J. (2011). A New 500m2 Paraboloidal Dish Solar Concentrator. *SolarPACES 2009*, 85(4), 620–626. https://doi.org/10.1016/j.solener.2010.01.009

Mawire, A., & Taole, S. H. (2014). Experimental Energy and Exergy Performance of a Solar Receiver for a Domestic Parabolic Dish Concentrator for Teaching Purposes. *Energy for Sustainable Development*, 19, 162–169. https://doi.org/10.1016/j.esd.2014.01.004, https://www.sciencedirect.com/science/article/pii/S0973082614000076

McConnell, R., & Fthenakis, V. (2012). Concentrated Photovoltaics. In V. Fthenakis (Ed.), *Third Generation Photovoltaics* (Ch. 7). IntechOpen. https://doi.org/10.5772/39245. https://cdn.intechopen.com/pdfs/32594/InTech-Concentrated_photovoltaics.pdf

McConnell, R., Symko-Davies, M., & Hayden, H. (2005). *International Conference on Solar Concentrators for the Generation of Electricity or Hydrogen: Book of Abstracts*. National Renewable Energy Laboratory, Golden, CO. www.osti.gov/biblio/15016069

Meusel, M., Adelhelm, R., Dimroth, F., Bett, A. W., & Warta, W. (2002). Spectral Mismatch Correction and Spectrometric Characterization of Monolithic III—V Multi-Junction Solar Cells. *Progress in Photovoltaics: Research and Applications*, 10(4), 243–255. https://doi.org/10.1002/pip.407

Muhammad-Sukki, F., Ramirez-Iniguez, R., McMeekin, S. G., Stewart, B. G., & Clive, B. (2010). Solar Concentrators. *International Journal of Applied Sciences (IJAS)*, 1(1), 1–15.

Muzathik, A., Nik, W., Samo, K., & Ibrahim, M. Z. (2010). Reference Solar Radiation Year and Some Climatology Aspects of East Coast of West Malaysia. *American Journal of Engineering and Applied Sciences*, 3, 293–299. https://doi.org/10.3844/ajeassp.2010.293.299

Ogunleye, I. O. (2011). Constraints to the Use of Solar Photovoltaic as a Sustainable Power Source in Nigeria. *American Journal of Scientific and Industrial Research*, 2, 11–16.

Osterwald, C. R., & Siefer, G. (2016). CPV Multijunction Solar Cell Characterization. In *Handbook of Concentrator Photovoltaic Technology* (pp. 589–614). https://doi.org/10.1002/9781118755655.ch10. Wiley Online Library. https://onlinelibrary.wiley.com/doi/book/10.1002/9781118755655

Philipps, S. P., Bett, A. W., Horowitz, K., & Kurtz, S. (2015). *Current Status of Concentrator Photovoltaic (CPV) Technology (NREL/TP—5J00–65130, 1351597; p. NREL/TP—5J00–65130, 1351597)*. https://doi.org/10.2172/1351597. https://www.nrel.gov/docs/fy16osti/65130.pdf

Sahu, S. K., K, A. S., & Natarajan, S. K. (2021). Design and Development of a Low-cost Solar Parabolic Dish Concentrator System with Manual Dual-axis Tracking. *International Journal of Energy Research*, 45(4), 6446–6456. https://doi.org/10.1002/er.6164

Sripadmanabhan Indira, S., Vaithilingam, C. A., Chong, K.-K., Saidur, R., Faizal, M., Abubakar, S., & Paiman, S. (2020). A Review on Various Configurations of Hybrid Concentrator Photovoltaic and Thermoelectric Generator System. *Solar Energy*, 201, 122–148. https://doi.org/10.1016/j.solener.2020.02.090

Wojtczuk, S., Chiu, P., Zhang, X., Pulver, D., Harris, C., & Siskavich, B. (2011). 42% 500X Bi-Facial Growth Concentrator Cells. *AIP Conference Proceedings*, 1407(1), 9–12. https://doi.org/10.1063/1.3658283

Zanganeh, G., Bader, R., Pedretti, A., Pedretti, M., & Steinfeld, A. (2012). A Solar Dish Concentrator Based on Ellipsoidal Polyester Membrane Facets. *Solar Energy*, 86(1), 40–47. https://doi.org/10.1016/j.solener.2011.09.001

Zhu, L., Mochizuki, T., Yoshita, M., Chen, S., Kim, C., Akiyama, H., & Kanemitsu, Y. (2016). Conversion Efficiency Limits and Bandgap Designs for Multi-Junction Solar Cells with Internal Radiative Efficiencies Below Unity. *Optics Express*, 24(10), A740–A751. https://doi.org/10.1364/OE.24.00A740

5 Modeling the Feasibility of Using Solar PV Technology for Residential Homes in Brunei

Saiful Azmi Husain, Nur Liyana Asnan,
Ten Ru Yuh, and Fredolin Tangang

CONTENTS

HARNESSING SOLAR PV TECHNOLOGY FOR FUTURE HOMES IN BRUNEI TOWARDS CLIMATE CHANGE RESILIENCE

5.1 INTRODUCTION

The burning of fossil fuel releases greenhouse gases (GHGs), which accumulate over Earth's atmosphere and trap heat from escaping, thus, making Earth warmer. Hence, this phenomenon known as global warming will consequently

DOI: 10.1201/9781003367819-5

lead to the increase of global air temperature and ocean temperature, which leads to disaster risk events, such as drought and flash floods. In the Brunei Darussalam Nationally Determined Contribution (NDC) for 2020 [1], Brunei is committed to reducing greenhouse gas (GHG) emissions by 20% relative to business-as-usual levels by 2030. One of the ten key strategies stated in the Brunei Darussalam National Climate Change Policy is to adopt renewable energy by increasing its total share by at least 30% of total energy capacity in the power generation mix [2]. With Brunei recording its warmest year in 2016 as well as experiencing rapid urbanization and industrialization, the average temperature has already increased in the year 2020 by 1.25°C and is expected to increase further, according to climate model projection [2]. One of the best ways to mitigate this is to deploy solar PV technology as part of the renewable energy strategy. It has been found in many literature reviews that using solar PV technology is safe, reliable, and low-maintenance, and provides green energy without on-site pollution or emission [3,4,5]. Brunei is considered to have the biggest solar farm demonstration project in Southeast Asia; namely, Tenaga Suria Brunei (TSB). The project evaluated and installed solar PV modules to discover the most convenient and outstanding achievement of solar PV technology for local meteorological conditions. There are six different types of modules: (i) mono-crystalline silicon (m-cSi), (ii) polycrystalline silicon (p-cSi), (iii) amorphous silicon (a-Si), (iv) multi-junction silicon (tandem), (v) copper-indium-selenium (CIS), and (vi) hetero-junction with an intrinsic thin layer (HIT) that have been installed with a nominal capacity of 200 kWp in which the solar farm covers an area of about 12,000 square meters with exactly 9,234 pieces of solar panels.

The main objectives of the studies are as follows:

(a) Discuss the feasibility of harnessing this solar PV technology for future homes in Brunei, to enhance climate resilience, using Photovoltaic Geographical Information System (PVGIS) modeling, by estimating the energy output and irradiation received by selecting specific types of solar panels for TSB on-site locations in Brunei as well as looking at rooftop solar PV design considerations.

(b) Discuss the limitations and challenges of this study and how to move forward.

This chapter is organized as follows. In Section 5.1, we give an overview of Brunei's commitment to reducing GHG emissions and implementing its solar farm project. Section 5.2 discusses solar photovoltaics (PV), its solar PV module classification, and their performance in Brunei. Section 5.3 is dedicated to the results and discussion of using Photovoltaic Geographical Information System (PVGIS) modeling, by estimating the energy output and irradiation received by selecting and comparing crystalline silicon and CIS solar panels for TSB on-site locations in Brunei as well as looking at rooftop solar PV design considerations. Finally, Section 5.4 presents the conclusion.

5.2   SOLAR PHOTOVOLTAICS (PV)

Solar photovoltaics (PV) or solar cells are semiconductor devices that convert sunlight into direct current (DC) electricity without any heat engine interfering. A PV power generation system is electrically configured into modules and arrays, which can be used to charge batteries, operate motors, and power any number of electrical loads. These systems are measured in peak kilowatts (kWp) [6].

Photovoltaic cells are connected electrically in series and/or parallel circuits to produce higher voltages, currents, and power levels. Photovoltaic panels include one or more PV modules assembled as a pre-wired, field-installable unit. A photovoltaic array is a complete power-generating unit consisting of any number of PV modules and panels.

The performance of PV modules and arrays are generally examined at standard test conditions (STC) and under controlled environments [7]. The PV system performance is sensitive to multiple parameters, such as temperature, humidity, and weather conditions [7]. Standard test conditions are defined by a module (cell) operating temperature of 25°C (77°F), incident solar irradiance level of 1,000 W/m^2, and air mass of 1.5 spectral distribution. Since these conditions are not always typical of how solar PV modules and arrays operate in the field, actual performance is usually 85% to 90% of the STC rating.

Solar photovoltaic systems are like any other electrical power generating systems, with the exception that only the equipment used is different from that used for conventional electromechanical generating systems. However, the principles of operating and interfacing with other electrical systems remain the same and are guided by a well-established body of electrical codes and standards.

Although a solar PV array produces power when exposed to sunlight, a number of other components are required to properly conduct, control, convert, distribute, and store the energy produced by the array. Depending on the functional and operational requirements of the system, the specific components required may include major components, such as a DC-AC power inverter, battery bank, system and battery controller, auxiliary energy sources, and sometimes specific electrical load (appliances). In addition, an assortment of a balance of system (BOS) hardware, including wiring, overcurrent, surge protection and disconnect devices, and other power processing equipment. Figure 5.1 shows a basic diagram of a photovoltaic system and the relationship among individual components.

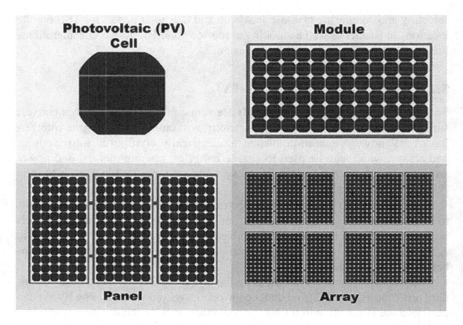

FIGURE 5.1　Photovoltaic cells, modules, panels, and arrays.

5.2.1　Classification of Solar PV Module

The major aspects influencing the selection of the world's solar cell materials are the minor variations of silicon purity, area, efficiency, and cost of the cell material [8]. Commercial manufacturing processes are generally used for further classification [8].

(i)　Monocrystalline silicon (m-cSi) solar cells [8]

Monocrystalline silicon solar panels are more aesthetic compared to polycrystalline silicon (p-cSi), as m-cSi possesses a more systematic look. The process to manufacture m-cSi is called Czochralski. M-cSi panels are capable of producing a power output of two to three times more than that of thin-film technology due to the space efficiency, in which the usual life span is about 25 years. As a result, m-cSi panels lead to more efficiency in warm weather and deteriorate in performance as temperature increases.

(ii)　Polycrystalline silicon (p-cSi) solar cells [8]

Polycrystalline silicon is a raw silicon that is melted and poured into a square mold, forming a perfectly looking rectangular shape without rounded edges.

P-cSi is simple and cheap. The wastage of silicon is lesser compared to m-cSi manufacturing. As a result, it is less heat tolerant and has lower space efficiency than m-cSi solar panels, obtaining a 13–16% efficiency and reduced life span.

(iii) Amorphous silicon (a-Si) solar cells [8]

Amorphous silicon is commonly used in small-scale applications, such as pocket calculators, as they produce low electrical power. A-cSi cells are manufactured by a process called stacking, where several layers of a-Si solar cells are combined, resulting in good efficiency, ranging between 6% and 8%. Generally, very low silicon, in the order of 1% of the silicon used in crystalline silicon solar cells, is used in a-Si solar cell manufacturing.

(iv) Copper-indium-selenide (CIS) solar cells [8]

Among the thin-film technologies, CIS solar cells possess good potential in terms of energy and efficiency in the range between 10% and 12%. The toxic nature of the CIS material presents a serious limitation on the technology.

(v) Multi-junction silicon (tandem) solar cells [9]

Multi-junction silicon (tandem) solar cells have the potential of achieving high conversion efficiencies of more than 40% and are promising for space and terrestrial applications.

(vi) Heterojunction with intrinsic thin layer (HIT) solar cells [10]

Heterojunction with intrinsic thin layer (HIT) solar cells have high PV conversion efficiencies and lower processing temperatures, which makes it attractive for large-scale commercialization to satisfy the demand for alternative sources to fossil fuel or nuclear energy.

5.2.2 Solar PV Performance Tested in Brunei

Brunei's first very own farm is capable of producing 1.2 MW peak capacity from six different types of solar modules: (i) m-cSi, (ii) p-cSi, (iii) a-Si, (iv) tandem, (v) CIS, and (vi) HIT [11,12].

According to Pacudan [12], Marion, Adelstein, and Boyle [13], there are three types of measure (metrics) used by the study to assess each module's performance:

(1) Yield factor Y(F) [12,13]

The yield factor refers to the plant's specific performance in net kWh delivered to the grid per kW of installed nominal PV module power.

$$Y(F) \quad \frac{E(real)}{P(STC)} \tag{5.1}$$

where:

$E_{(real)}$—Actual yield produced by the module and sent to the grid (kWh)

$P_{(STC)}$—Module output power at standard test condition (kW)

(2) Performance ratio [12,13]

Performance ratio (*PR*) measures how the solar PV plant effectively converts sunlight into AC energy delivered to the off-taker relative to what would be expected from the module's rated capacity. A dimensionless indicator refers to the actual amount of solar power produced by a module in comparison to the maximum possible power output of the module.

$$PR = \frac{Effective\ yield\,(AC)}{Maximum\ theoretical\ yield\,(DC)} \tag{5.2}$$

The previous formula can be simplified and calculated as follows:

$$PR = \frac{E(x)}{I(solar) x P(STC)\,/\,I(STC)} \tag{5.3}$$

where:

$E_{(x)}$—Actual yield (theoretical yield) produced by the module and sent to the grid (kWh)

$I_{(solar)}$—Actual total solar irradiation (kWh/m²)

$P_{(STC)}$—Module output power at standard test condition (kW)

$I_{(STC)}$—Standard test condition irradiation (kWh/m²)

(3) Spatial yield Y(A) [12,13]

Spatial yield *Y(A)* measures the energy yield that each module type can generate in a given land area (expressed in kWh/m²):

$$Y(A) = \frac{E(real)}{A(p)} \tag{5.4}$$

where:

$E_{(real)}$—Actual yield produced by the module and sent to the grid (kWh)

A_p—Solar PV power plant land area requirement

TABLE 5.1

Module Performance Ranking Based on Yield Factor and Performance Ratio [12]

Ranking	Module	Specific Yield (kWh/kW)	Performance Ratio
1	a-Si	1,501	0.77
2	HIT	1,475	0.75
3	CIS	1,403	0.72
4	p-cSi	1,384	0.71
5	Tandem	1,370	0.70
6	m-cSi	1,330	0.68

Yield factor: Amorphous silicon achieved a higher yield factor (1,501 kWh/kW), with HIT coming second with 1,475 kWh/kW and followed by CIS with 1,403 kWh/kW (Table 5.1). A specific yield of more than 1,400 kWh/kW is already considered high for the Southeast Asian region standards.

For the performance ratio: Amorphous silicon recorded the highest value, with 0.77, while mono-crystalline silicon recorded the lowest (0.68). It is stated that currently, newly manufactured solar PV modules have a performance ratio of more than 0.90, and those values below 0.85 are already considered poor.

For the spatial yield performance, HIT provides the highest yield per unit area. Amorphous silicon has the lowest spatial yield. This performance indicator is influenced mainly by module efficiency. The module with a higher efficiency would result in a higher spatial yield. The specifications of these systems are given in Table 5.2. As what has been stated, these modules have different conversion efficiencies, and hence, the number of modules and the panel area of 200 kWp rated capacity are different. Table 5.3 shows the array parameters of each module given during the TSB project.

TABLE 5.2

Module Performance Ranking Based on Spatial Yield [12]

Ranking	Module	Module Efficiency (%)	Spatial Yield (kWh/m²)
1	HIT	16.0	235.0
2	p-cSi	13.4	182.3
3	m-cSi	14.1	182.0
4	CIS	8.8	142.0
5	Tandem	8.2	113.0
6	a-Si	6.3	95.0

TABLE 5.3

Module Types and Array Parameters of the TSB Project [12]

	Type 1	Type 2	Type 3	Type 4	Type 5	Type 6
Type of cell	m-cSi	p-cSi	Tandem	a-Si	CIS	HIT
Manufacturer	Sharp	Mitsubishi Electric	Mitsubishi Heavy Industries	Mitsubishi Heavy Industries	Solar Frontier	Sanyo
Model	NU-S0E3E	PV-TD185MF5	MT-130	MA100T2	SC80EX-A/B	HIP-205NKHB5
Module efficiency	14.1%	13.4%	8.2%	6.3%	8.8%	16.0%
Maximum power (Pmax)	180	185	130	100	80	205
No. of modules for 200 kW	1,116	1,098	1,540	2,000	2,500	980
Total area for 200 kW	1,462	1,518	2,426	3,150	1,979	1,257
No. of array	62	61	129	100	125	98

5.3   RESEARCH BY PVGIS-SARAH

By studying the performance of grid-connected PV (Figure 5.2), PVGIS-SARAH from Photovoltaic Geographical Information Systems (PVGIS) [14] is being used to estimate solar electricity generation. The PVGIS-SARAH data set is based on a new algorithm developed by the Climate Monitoring Satellite Application Facility (CMSAF). PVGIS adjusts the database every few years to suit changing solar irradiation values and makes the data record more up-to-date. The study only examines the result between (i) crystalline silicon and (ii) CIS. The following is the result of TSB's PV performance:

   (1) Crystalline silicon

5.3.1   Statistical Measurements for the Quality Evaluation of the Solar Surface Radiation Data Sets

The verification operates on several statistical measures and results to evaluate the quality of the solar surface radiation data sets (Table 5.4 and Figure 5.3, and Table 5.5 and Figure 5.4) [16]. According to Richard et al., the surface measurements and the CMSAF data set, commonly calculated using bias and standard deviation (SD), can be also calculated by the mean absolute differences (or mean absolute bias, MAB) and the anomaly correlation (AC) [16].

   The formulas of the statistical measures are taken from [16,17]. In the following, by definition, the variable y describes the validated data set (e.g., PVGIS-SARAH), and $\bar{o}$ denotes the reference data set (e.g., Baseline Surface Radiation Network (BSRN)). The individual time step denoted with k and n is the total number of time steps.

   (i) Bias [16]

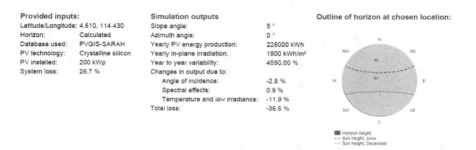

FIGURE 5.2   Performance of grid-connected PV at TSB [15].

TABLE 5.4

Monthly PV Energy and Solar Irradiation [15]

Month	Average Monthly Electricity Production (kWh)	Average Monthly Sum of Global Irradiation by the Modules (kWhm^{-2})	Monthly Standard Deviation Electricity Production— Year-to-Year Variation (kWh)
January	18,500	143	2,310
February	19,200	150	1,960
March	22,200	174	1,530
April	20,000	160	1,030
May	18,900	150	879
June	17,400	138	848
July	18,600	148	564
August	19,500	154	682
September	18,500	146	1,040
October	18,500	145	1,170
November	18,200	144	1,390
December	18,600	145	1,340

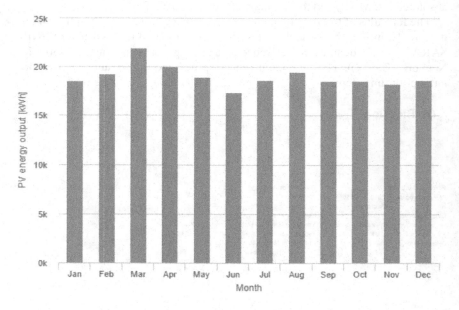

FIGURE 5.3 Monthly energy output from fix-angle PV system (generated on 2019/04/07) [15].

The bias is the mean difference between the two considered data sets. It indicates whether the data set on average is over or underestimates the reference data set.

$$Bias = \frac{1}{n}\sum_{k=1}^{n}(y_k - o_k) - \bar{y} - \bar{o} \tag{5.5}$$

(ii) Standard deviation (SD) [16]

SD is a measure of the spread around the mean value of the distribution formed by the differences between the generated and the reference data set.

$$SD = \sqrt{\frac{1}{n}\sum_{k=1}^{n}(y_k - o_k) - (\bar{y} - \bar{o})^2} \tag{5.6}$$

(iii) Mean absolute bias (MAB) [16]

In contrast to the bias, the MAB is the average of the absolute values of the differences between each member of the time series. The advantage of the MAB is that there is no cancellation of positive and negative values.

$$MAB = \frac{1}{n}\sum_{k=1}^{n}|y_k - o_k| \tag{5.7}$$

(iv) Anomaly correlation (AC) [16]

AC describes to which extent the anomalies of the two considered time series correspond to each other without the influences of a possibly existing bias. The AC retrieved from satellite data and derived from surface measurements allows the estimate of the potential to determine anomalies from satellite observations.

$$AC = \frac{\sum_{k=1}^{n}(y_k - \bar{y})(o_k - \bar{o})}{\sqrt{\sum_{k=1}^{n}(y_k - \bar{y})^2}\sqrt{\sum_{k=1}^{n}(o_k - \bar{o})^2}} \tag{5.8}$$

(2) CIS

As a result, by comparing crystalline silicon and CIS, it shows that the average monthly electricity production of crystalline silicon produced more energy than CIS, with an average estimate of 18,990 kWh in one year (Figure 5.5).

5.3.2 POWER GENERATION

The power generation for the PV system can be classified into two ways of power generation:

(i) Photovoltaic power generation

TABLE 5.5

Monthly PV Energy and Solar Irradiation (Generated on 2019/04/07) [15]

Month	Average Monthly Electricity Production (kWh)	Average Monthly Sum of Global Irradiation by the Modules (kWhm^{-2})	Monthly Standard Deviation Electricity Production—Year-to-Year Variation (kWh)
January	18,100	143	2,330
February	18,900	150	1,980
March	21,700	174	1,540
April	19,900	160	1,050
May	18,700	150	892
June	17,200	138	860
July	18,500	148	572
August	19,300	154	695
September	18,200	146	1,050
October	18,100	145	1,190
November	18,000	144	1,420
December	18,300	145	1,350

Provided inputs:

Latitude/Longitude:	4.610, 114.430
Horizon:	Calculated
Database used:	PVGIS-SARAH
PV technology:	CIS
PV installed:	200 kWp
System loss:	26.7 %

Simulation outputs

Slope angle:	5 °
Azimuth angle:	0 °
Yearly PV energy production:	225000 kWh
Yearly in-plane irradiation:	1800 kWh/m²
Year to year variability:	4660.00 %
Changes in output due to:	
Angle of incidence:	-2.8 %
Spectral effects:	? (0) %
Temperature and low irradiance:	-12.2 %
Total loss:	-37.4 %

Outline of horizon at chosen location:

FIGURE 5.4   Performance of grid-connected PV at TSB (generated on 2019/04/07) [15].

A grid-connected system is connected to a large independent grid, which, in most cases, is the public electricity and feeds power into the grid [6]. There are varying sizes, from a few kWp for residential purposes to solar stations up to tens of GWp, to form a decentralized electricity generation [6]. Poponi stated that the prospects for the diffusion of electricity generation from PV technology in grid-connected systems by the experience methodology curves are used to predict the different levels of cumulative world PV shipments required to reach the calculated break-even PV systems prices, assuming different trends in the relationship between price and the increase in cumulative shipments [18].

(ii) Hybrid photovoltaic power generation

The hybrid power generation system combines a renewable energy source of PV with other forms of generation, usually a conventional generator powered

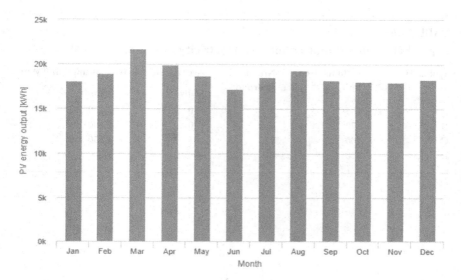

FIGURE 5.5 Monthly energy output from fix-angle PV system (generated on 2019/04/07) [15].

by diesel or even another renewable form of energy [6]. Hybrid systems serve to reduce the consumption of non-renewable fuel [6]. A method of modeling an energy store was used to match the power output from a wind turbine and a solar PV array to a varying electrical load, and the method was validated against time-stepping methods, showing good agreement over a wide range of store power ratings, store efficiencies, wind turbine capacities, and solar PV capacities [19].

5.3.3 CASE STUDY IN BRUNEI

5.3.3.1 Typical Household Electricity Consumption Profile

According to Pacudan [12], the analysis of electricity consumption pattern in Brunei depends on the typical residential house (Table 5.6 and Figure 5.6); for example, the following is in Rimba:

   (i) Type A: Terrace house
  (ii) Type B: Detached house
 (iii) Type C: Single average house
 (iv) Type D: Single big house

With the average temperature already increasing since the year 2020 by 1.25°C [2] and is expected to increase further according to climate model projections,

TABLE 5.6

Typical Household Information and Electricity Consumption [20]

Typical House	Building Area (m²)	No. of Bedrooms	Average Electricity Consumption (kWh)	
			Daily	Month
A	135	4	50	1,500
B	270	6	70	2,100
C	400	7	85	2,550
D	550	8	100	3,000

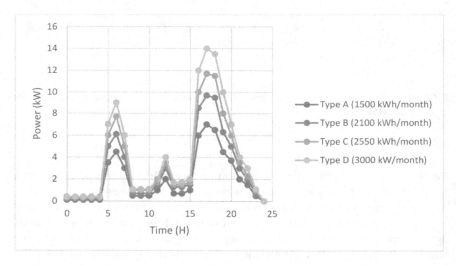

FIGURE 5.6 Daily consumption profile for typical households [20].

in addition to rapid urbanization and industrialization, solar PV technology installation is the way forward. But solar PV technology needs more space for installation. Currently, the main issues are the typical house's total rooftop area limitation and availability of free space surrounding the house for solar PV to be installed on the ground level.

5.3.3.1.1 Roof Area Limitation

Electricity generation and PV plant capacity installation is, however, constrained by the area available on the rooftop of the house [20].

Example:

As stated by Pacudan [20], the 10 kWp solar PV plant used in the analysis requires a total roof area of 65 m² (Figure 5.7). Therefore, this implies that Types A and B houses may able to provide a smaller system according to the roof area,

FIGURE 5.7 Roof area required for 10 kWp solar PV plant [20].

and that only Types C and D houses may have sufficient roof spaces for a 10 kWp PV system.

5.3.3.1.2 Opportunity Cost of Consumed Electricity

The solar PV power plant will displace grid generation; hence, the opportunity cost of consuming power from PV installation is represented by the price of the grid power [20], which is more expensive than the conventional new electricity tariff currently imposed in Brunei (Table 5.7). This is due to the blessing of having a subsidized electricity rate in Brunei.

TABLE 5.7

Illustration of an Opportunity Cost of Consumed Electricity for House Type D [20]

Units (kWh)	Tariff Rate (BND/kWh)	Electricity Consumption of House Type D (kWh/month)	Opportunity Cost (B$)
1–600	0.01	600	6.00
601–1,500	0.08	899	71.92
1,501–4,000	0.10	1,501	150.10
4,001 and above	0.12		
Total		3,000	228.02
Average			B$ cents 7.6/kWh

5.3.4 Discussion

Solar PV systems are currently a popular trend and beneficial to average households across Australia [3]. Many households in Australia are enjoying the advantages of installing solar to reduce electricity bills and have less reliance on the grid, just to name a few [3]. The following are some of the advantages [20]:

(i) Environmentally friendly

Solar energy is a great alternative to traditional fossil fuels, utilizing the free light energy created by the sun. Solar also does not release any harmful greenhouse gases, like carbon dioxide, sulfur dioxide, ammonia, nitrogen oxides, and particulate matter, which are emitted when burning coal, and, thus, are associated with traditional coal fire power stations. Moreover, solar energy does not produce radioactive waste since it is renewable and freely available.

(ii) Protection from electricity price rises

Due to the fact that electricity prices steadily increase each year, having a solar system helps to limit the purchase amount of electricity from the grid and protects from electricity price rises and fluctuations.

(iii) Accessibility

Solar energy will always be freely available as long as the sun exists. As the source of solar energy systems becomes more popular, more installers and providers become Clean Energy Council accredited, making it easier and more accessible to find quality providers for the next five billion years.

(iv) Low maintenance

As there are no moving parts, solar PV systems require very little to no maintenance. Most solar panels have a life span of around 20 years and come with warranties for the majority of the time. Cleaning is recommended every 6–18 months, depending on the location.

5.3.5 Results

According to Pacudan, the TSB project has assessed the performance of the six different solar PV modules based on the calculated data using the metrics method [12]. Hamidah et al. [7] also assessed and compared the performance of the six different solar PV modules following the International Energy Agency (IEA) guidelines, taking the array yield, capture losses, array efficiency ratio, and performance ratio as the criteria under the tropical environment. As shown on Table 5.1–5.3, it can be concluded that a-Si, HIT, and CIS have recorded preferred results compared to m-cSi and p-cSi modules due to the yield factor and performance ratio [12,7]. However, HIT, m-cSi, and p-cSi modules also show a better energy yield per unit area and give a result that a-Si generates the least output per unit area [12,7].

5.4 CONCLUSION

The advantage of solar PV systems is that they are much cheaper, can be built in any size required, and also offer more possibilities of use, and have less stringent maintenance demands. Moreover, solar PV systems cover part of the electricity demand; the excess power can be fed into the utility grid (grid-connected system) and stored in the battery system for later use or for heating (for example, electric water heater).

Solar energy's contribution to the total global energy is very low, but the potential is enormous, and this provides security against conventional fuel supply disruptions and their prices. The way to move forward in terms of harnessing this solar PV technology for future homes in Brunei, in enhancing climate change resilience, is by planning and developing a smart city, in which all green residential homes or buildings' rooftops are specially designed for solar PV installation with optimal solar irradiance and energy output performance via modeling software and using other technology (for example, drone driven for solar panel cleaning, automated solar tracking system, and seasonal tilt system). Grid-connected PV power generation is probably the best way to implement in typical residential homes for sustainable electricity generation since solar PV can be used for lights and generate electricity for small electrical appliances, and electrical heating systems, whereas the utility grid could be used to generate electricity for heavy-duty electrical appliances, such as air conditioning, even though there are technical issues and limitations as well as hidden costs associated with managing these solar PV systems. Attractive electricity price plans and packages should be developed in order to establish better electricity tariff rates for these typical green residential smart homes, taking into account daily and seasonal weather variations and when the peak demands mostly occur in the late afternoon to early evening.

REFERENCES

[1] Brunei Darussalam First National Determined Contributions (NDCs) (2020). Retrieved from https://unfccc.int/documents/497350

[2] Brunei Darussalam National Council of Climate Change (2020). Retrieved from www.climatechange.gov.bn/SitePages/BNCCP/index.html#page=2

[3] Advantages of Solar PV. (n.d.). Retrieved from www.solarmarket.com.au/residential-solar/advantages-of-solar-pv/

[4] Kala Meah et al. (2008). Solar Photovoltaic Water Pumping for Remote Locations. *Renewable and Sustainable Energy Reviews*, 12(2), 472–487.

[5] Kumar, B. S., & Sudhakar, K. (2015). Performance Evaluation of 10 MW Grid Connected Solar Photovoltaic Power Plant in India. *Energy Reports*, 1, 184–192.

[6] Parida, B., Iniyan, S., & Goic, R. (2011). *Renewable and Sustainable Energy Reviews: A Review of Solar Photovoltaic Technologies.* Mechanical Engineering and Naval Architecture University of Split, Croatia, and Anna University Chennai, Chennai, India.

[7] Hamidah, I., et al. (2013). Comparative Performance of Grid Integrated Solar Photovoltaic Systems Under the Tropical Environment. *IEEE Innovative Smart Grid Technologies-Asia (ISGT Asia),* pp. 1–6. https://ieeexplore.ieee.org/document/6698758

[8] Hudedmani, G., Sopimath, V., & Jambotkar, C. (2017). *A Study of Materials for Solar PV Technology and Challenges.* KLE Institute of Technology, Gokul, Hubbali, Karnataka, India.

[9] Yamaguchi, M. (2003). *III—V Compound Multi-Junction Solar Cells Present and Future.* Toyota Technological Institute, 2–12–1 Hisakata, Tempaku, Japan.

[10] Iftiquar, S. M., Youngseok, L., et al. (2013). *High Efficiency Heterojunction with Intrinsic Thin Layer Solar Cell: A Short Review.* College of Information and Communication, Sungkyunkwan University, Korea & Department of Energy Science, Sungkyunkwan University, Korea.

[11] Tenaga Suria Brunei, Largest Solar Power Plant (2010). Retrieved from https://thailand.prd.go.th/ewt_news.php?nid=257&filename=index

[12] Pacudan, R. (2015). *1.2 MWp Tenaga Suria Brunei Solar PV Power Generation Demonstration Project.* Brunei National Energy Research Institute, Brunei.

[13] Marion, B., Adelstein, J., et al. (2005). *Performance Parameters for Grid-Connected PV Systems.* National Energy Laboratory, NREL/CP-520–37358. Golden, Colorado USA.

[14] Photovoltaic Geographical Information Systems (PVGIS). Retrieved from https://re.jrc.ec.europa.eu/pvg_tools/en/tools.html#PVP

[15] Performance of Grid-Connected PV. (2017). *PVGIS European Union* (Generate on 2019/04/07). https://joint-research-centre.ec.europa.eu/pvgis-online-tool_en

[16] Richard, M., Uwe, P., et al. (2015). *Digging the METEOSAT Treasure: 3 Decades of Solar Surface Radiation.* Deutscher Wetterdienst, Frankfurter Str. 135, D-60387 Offenbach, Germany.

[17] Wilks, D.S. (2006). *Statistical Methods in the Atmospheric Sciences.* Academic Press, New York.

[18] Poponi, D. (2003). *Solar Energy: Analysis of Diffusion Paths for Photovoltaic Technology Based on Experience Curves.* https://www.sciencedirect.com/science/article/abs/pii/S0038092X03001518

[19] John, P.B., & David, G.L. (2006). A Probabilistic Method for Calculating the Usefulness of a Store with Finite Energy Capacity for Smoothing Electricity Generation from Wind and Solar Power. *Journal of Power Sources,* 162, 943–948.

[20] Pacudan, R. (2014). *Residential Solar PV Policy FEED-IN TARIFF vs NET METERING: Options for Brunei.* Brunei National Energy Research Institute, Brunei.

6 Synthesis Techniques and Applications of Graphene and Its Derivatives

Hamidatu Alhassan, Yvonne Soon Ying Woan,
Anwar Usman, and Voo Nyuk Yoong

CONTENTS

6.1 INTRODUCTION

The COVID-19 crisis influenced the highest dependency of world economies on coal to power economic growth in 2021, resulting in the worst CO_2 emissions recorded in history (European Commission, 2022; International Energy Agency, 2022). From 34.81 billion tonnes in 2020, it rose by 6% to 36.3 billion tonnes. With the current global shortage in gas supply, most countries, especially import-dependent countries, are in economic turmoil as fuel prices wreak havoc (World Economic Forum, 2022). A pragmatic way to tackle the situation and prevent future crises is by considering alternative sources of energy generation.

Fossil energy sources, such as coal, natural gas, petroleum, and oil, account for about 80% of global primary energy consumption (Abas et al., 2015). The burning of fossil fuels releases CO_2 and other greenhouse gases into the atmosphere, which poses a threat to the climate (Hegerl et al., 2019). The risk of depletion of fossil fuels further reinforces the demand for more sustainable methods.

DOI: 10.1201/9781003367819-6

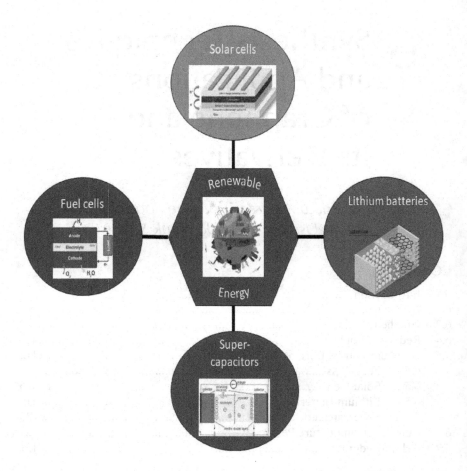

FIGURE 6.1   Energy conversion and storage devices for renewable energy generation.

However, energy generated through renewable sources, such as solar and wind energy, requires a converting device and storage system to circumvent the issue of intermittency (Yin et al., 2020). Scientists are striving to seek and design materials suitable for energy storage and conversion technology generally found in fuel cells, solar cells, supercapacitors, and batteries, as shown in Figure 6.1 (Salvia & Brandli, 2020). The most advanced materials for energy storage and conversion technology applications are graphene and its derivatives, specifically graphene oxide (GO) and reduced graphene oxide (rGO) (see Figure 6.2).

Graphene is a two-dimensional one-atom-thick material that exists as a building block of graphite (Shams et al., 2015). It contains six sp^2-bonded carbon atoms tightly arranged in a honeycomb lattice (Shams et al., 2015). The term "graphene" was used to depict single graphite sheets for the first time in 1987 (Głuchowski

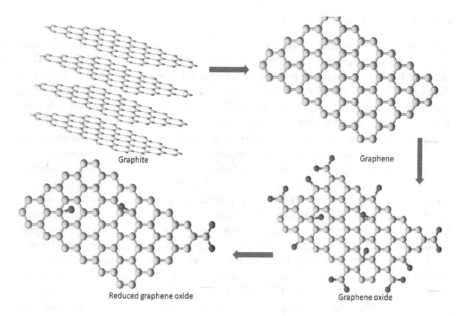

Graphite

Graphene

Reduced graphene oxide

Graphene oxide

FIGURE 6.2 Graphene and its derivatives.

et al., 2020). Graphene has displayed the most remarkable properties among all sp^2 carbon allotropes. It is the strongest, thinnest, and lightest material known to man, and the best electricity and heat conductor ever discovered. The remarkable properties of graphene in terms of solar radiation absorptivity, mechanical stability, thermal and electrical conductivity, surface area, theoretical capacitance, and ultra-thinness are harnessed for energy storage and conversion technology. Ease of functionalization to tune the electrical, optical, and electrocatalytic properties gives it good prospects in energy harvesting applications. Many research works have been carried out on incorporating graphene as additives or standalone materials in components of energy conversion and storage devices to enhance power output and stability.

In this chapter, the role of graphene and its derivatives in solar cells, fuel cells, supercapacitors, and lithium batteries are summarized. Firstly, the properties and current synthesis techniques of graphene and its derivatives are discussed. The viability of the synthesis methods and their derivatives are assessed, and the challenges that inhibit their scale-up processes are identified. An overview of various top-down and bottom-up approaches is given and evaluated against variables such as product quality, scalability, cost, and environmental impact. Secondly, we discuss and elaborate on the potential utilization of graphene properties in improving the performance of energy conversion and storage devices. Finally, the key findings are summarized, and the possible directions for future research are proposed.

6.2 GRAPHENE

Graphene is considered a wonder material due to its multitude of unique and favorable properties resulting from the substantial π-electron conjugation (Liu et al., 2019). It is recognized as the thinnest material, with an atomic thickness of 0.345 nm (one-atom thick). It is the toughest 2D material (200 times stronger than steel) with a tensile strength of over 1 TPa (X. J. Lee et al., 2019). Though very thin, it is capable of absorbing 2.3% of white light, which makes it very visible to the human eye (Hayes, 2019). It is an isotropic conductor with very high thermal conductivity of 5,000 W/m-K (higher than diamond) and high current mobility of 250,000 cm^2/Vs (X. J. Lee et al., 2019). However, as the most efficient conductor in existence, graphene applications in electronics are limited by the absence of a bandgap. To eliminate this limitation, experimental procedures, such as quantum confinement, substitutional doping, graphene hybrids, and substrate-induced bandgap opening, have been widely examined (Jariwala et al., 2011). These procedures result in more active sites for stronger molecular adsorption and changes in electronic features, which is relevant for electrocatalytic reactions. Graphene can serve as a microwave receiver for signals up to 2.45 GHz and has a single-layer transmittance as high as 97.7% (Bhuyan et al., 2016). It has a light weight of about 0.77 mg/m^2 and the largest surface area of 2,630 m^2/g among all materials (Bhuyan et al., 2016). The elasticity of graphene is such that it can be stretched to about 20% of its initial size without reaching its breaking point (Bhuyan et al., 2016). It has an extremely low resistivity (lower than silver) at room temperature and a low gas permeation even to helium atoms (Y.-C. Wang & Cho, 2017).

Graphene is an inert material, but its chemical functionality can be changed to improve its properties (*Wet Chemistry of Graphene*, n.d.). It can be functionalized with fluorine to produce fluorinated graphene or be introduced with C-O covalent bonds to produce graphene oxide (Dimiev & Tour, 2014). These remarkable properties and super flexibility have made graphene the doctors', engineers', researchers', and environmentalists' favorite. However, most of these features are only associated with graphene that has no defects (pristine graphene). Defects, such as interstitial impurities, heterostructures, vacancies, edges, and grain boundaries, in a graphene material can arise from the experimental preparation processes (Banhart et al., 2011). The presence of defects can either prove favorable or detrimental to the function of the material depending on the intended application (Xu et al., 2012). The deliberate insertion of defects can induce new functionalities. The molecular dynamics simulation results by Jing et al. showed that the existence of Stone-Wales (SW) defects and vacancy defects reduces the Young's modulus of graphene (Jing et al., 2012). However, their reconstruction consequently aided in stabilizing the Young's modulus. By functionalizing the vacancy-defective graphene with hydrogen atoms, an increase in the Young's modulus was observed.

The first monolayer of graphene sheets was unambiguously isolated in 2004 through the Scotch tape technique (Geim & Novoselov, 2009). One-atom-thick crystallites were extricated from graphite by Andre Geim and K. Novoselov.

Using adhesive tapes, graphene layers were peeled off from layers of highly oriented pyrolytic graphite (HOPG) repeatedly until there was a single layer left. It was then transferred onto a thin silicon dioxide (SiO_2) substrate where further processing can be performed. Although simple and straightforward, the graphene produced is highly defective, with irregular shapes and orientation, reflected by a relatively low electrical conductivity of 1,820 S/cm (Lin et al., 2013).

Inspired by the Scotch tape technique is the three-roll mill machine technique, which uses polymer adhesives as micromechanical cleavage to produce 1.13–1.14 nm thick graphene sheets (Chen et al., 2012). The polymer adhesive is made up of polyvinyl chloride (PVC) dissolved in dioctyl phthalate (DOP). Relative to the Scotch tape method, this technique has been proven feasible for large-scale production and is widely applied in the rubber industry. It is mainly limited by adhesive cost and energy requirement for the purification of graphene sheets from leftover PVC and DOP (Chen et al., 2012). Different routes of synthesis generally classified into top-down and bottom-up approaches have since been explored, as schematically illustrated in Figure 6.3. In Table 6.1 and Table 6.2, the properties and advantages of graphene produced by some top-down and bottom-up approaches, and the limitations of fabrication are summarized, respectively.

Following its discovery, graphite had been the main carbon source utilized in synthesizing graphene through a layer-by-layer exfoliation of the compound until a honeycomb carbon sheet remains. In the same year of its successful isolation, a group of researchers managed to use a non-graphitic source to produce graphene; epitaxial growth via thermal decomposition of silicon carbide (SiC) on the (0001) surface (Berger et al., 2004). Subsequently, ruthenium (Ru) was utilized as a substrate for the epitaxial growth of graphene (Sutter et al., 2008).

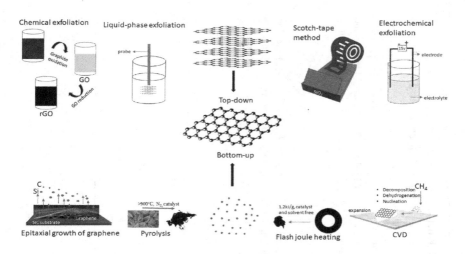

FIGURE 6.3 Bottom-up and top-down synthesis approaches.

TABLE 6.1

Overview of Some Properties and Challenges for Fabricating Top-Down Graphene

Top-Down Method	Size	Number of Layers	Electronic Properties	Advantages	Challenges for Fabrication	Reference
Micromechanical exfoliation	1.13–1.41 nm thickness	Monolayer	Electrical conductivity = 277 S/cm	• Low cost of production • Defect-free • High quality	• Low yield • Challenging to identify the number of layers • Difficult to remove residual polymer adhesives	(Lin, Chen et al., 2013; Yi & Shen, 2015)
Electrochemical exfoliation	44 µm diameter	Stacked multilayers (≤ 3)	• Hole mobility = 310 cm^2/Vs • Sheet resistance = 2 kΩ/sq	• Low defect content • Single-step process • Low operation time • High yield • Environmentally friendly	• High energy requirement • Expensive • Inevitable overoxidation of graphite with acidic electrolytes • Disruption of electronic properties with ionic electrolyte	(Parvez et al., 2014; Tiwari et al., 2020)
Ultrasound-assisted liquid-phase exfoliation	3 nm thickness, 1–2 µm diameter	Few layers (< 10)	Electrical conductivity = 1.72×10^{-4} S/m	• Simple operation process • Economically benign • Excellent dispersion stability	• Multistep post-synthesis processes • Inducible to edge-type defects • Low yield • High energy requirement • Time-consuming • Difficult to remove impurities	(Y. Oh et al., 2013; Sadhukhan et al., 2016)

TABLE 6.2

Overview of Some Properties and Challenges for Fabricating Bottom-Up Graphene

Bottom-Up Method	Size	Number of Layers	Electronic Properties	Advantages	Limitations of Fabrication	Reference
Chemical vapor deposition	4.1 μm diameter 1.5 nm thickness	Few layers of graphene	Sheet resistance = 1.9 K Ω/sq; Electron mobility = 1.5×10^4 cm²/V s	• Large area • High quality • Relatively high purity • Impervious	• Transfer process may induce defects and impurities • Low yield • Release of hydrogen • Complex and long operation process • High temperature • Expensive	(Burgess et al., 2011; Obraztsov et al., 2007)
Epitaxial growth	5 μm diameter 0.8 nm thickness	Monolayer and bilayer graphene	Electron mobility = 104–304 cm²/Vs	• Patterned graphene structure • High quality • Direct growth of graphene	• Low yield • Relatively small wafer size • High cost of SiC substrate	(Riedl et al., 2010; Sprinkle et al., 2010)
Pyrolysis	1–5 nm sheet thickness	Few layers (<10)	Sheet resistance = 1.5–3 KΩ/sq. Conductivity = 470 S/m	• Rapid processing • Cost effective • Ample quantity	• Relatively low quality • Defective graphene	(Raghavan et al., 2017; Zhao & Zhao, 2013)
Flash joule heating	~1 nm thickness	3–8 layers	-	• High quality • Relatively low defects • Turbostratic graphene sheets • Cheap • Fastest route • Environmentally friendly • Reduces carbon footprint	• Limited research on its application • Limited research on electronic properties	(W. Chen et al., 2022; Qiu et al., 2022)

The CVD technique, employing gaseous hydrocarbons, such as methane, followed shortly after the layer-by-layer exfoliation (Q. Yu et al., 2008). Solid hydrocarbons, such as SiC, were later introduced to the CVD method (Hofrichter et al., 2010). Since then, multiple carbon sources, including liquid, gaseous, and solid hydrocarbons, have been employed in the CVD graphene growth process. Amidst the performance of many bottom-up and top-down approaches in producing graphene of high quality, CVD is identified as the most promising and popular technique due to its potential for scalability (Saeed et al., 2020). The first successful CVD synthesis of few-layered graphene was in 2006 (Somani et al., 2006). In the CVD process, gas molecules are combined in a chamber reaction set at ambient temperature (1,000–1,200°C). Upon contact with a substrate, a reaction occurs that forms a material on the surface of the substrate, and the chamber is set to cool. The type of reaction is primarily dependent on the choice of substrate and the temperature of the substrate (Dasari et al., 2017). The substrate functions as the reaction catalyst. The resulting graphene is of high purity, has increased hardness, has a large surface area, and has high electron mobility (X. Li et al., 2009).

The most infant route for graphene synthesis is flash joule heating, which was introduced for the first time by Tour and his colleagues at Rice University in 2020 (Algozeeb et al., 2020). The carbon atoms in any item containing carbon, such as food waste, plastic water bottles, and worn-out rubber tires, will be reordered to produce graphene when a split-second super-hot flash of electricity is focused on it (Stanford et al., 2020). This process will break all the chemical bonds and cause all other elements to evaporate as gases. By placing any solid carbon material between two electrodes and applying 200 V of the short electrical pulse, the temperature of the carbon material instantaneously hits about 3,000°C. Every chemical bond in the material breaks. Non-carbon elements sublime out, and the remaining carbon atoms reconstruct into one-atom-thick turbostratic graphene with misaligned sheets.

6.3   GRAPHENE OXIDE

Graphene oxide (GO) is a monolayer of carbon similar in structure to graphene with a disrupted sp^2 bonding network and thickness of around 1 nm (Tarcan et al., 2020). Both sides are modified with oxygen-containing functional groups, such as epoxy, carboxyl, alkoxy, and hydroxyl, attached to the borders of the sheet. The presence of oxygenated moieties facilitates covalent interaction with other molecules and the formation of stable aqueous dispersions in a wide range of organic solvents, increased interlayer distance, and colloidal stability. Graphene oxide is mainly obtained by chemical oxidization and exfoliation of graphite.

The synthesis of graphene oxide dates back to 1859, when British chemist Benjamin Brodie investigated the reactivity of graphite flakes using potassium chlorate as an oxidizing agent (Brodie, 1860). The dispersibility of the resulting GO was limited to pure water. In 1898, L. Staudenmaier improved Brodie's method by introducing concentrated sulfuric acid and the gradual addition of potassium chlorate. The reaction was time-consuming and highly toxic due to the emission of NO_2, N_2O_4, and ClO_2. In later years, Hummers and Offeman developed a different recipe that replaced potassium chlorate with potassium permanganate (Hummers & Offeman, 1958). With Hummer's method, the reaction time was relatively short (2 h) and produced GO with a higher degree of oxidation.

Currently popular in the research community is the Macarno-Tour method, also known as the improved Hummer's method (Marcano et al., 2010). They increased the amount of oxidant and reduced reaction toxicity by eliminating sodium nitrate. The resulting GO was of high oxidation degree and had a higher yield of well-oxidized hydrophilic carbon material up to 97%. Unlike graphene, GO does not absorb visible light and has a relatively low electric conductance, causing a semi-conductive or insulative behavior depending on the degree of oxidation (Tian et al., 2021). In the literature, it is reported to have a sheet resistance of ~1,010 Ω/Sq and conductivity of 3.86×10^{-9} S/cm (Zainal et al., 2020).

Graphene oxide was initially regarded as the result of only oxidizing and exfoliating graphite (Dideikin & Vul', 2019). However, recent experimental results have demonstrated that GO can also be obtained by hydrothermal treatment of glucose and chemical vapor deposition (CVD) techniques (L. Sun, 2019).

6.4   REDUCED GRAPHENE OXIDE

Thermal, chemical, and photothermal processes are employed to reduce graphene oxide as an attractive route to produce functionalized graphene or reduced graphene (Dideikin & Vul', 2019). The reduction process removes the oxygen-containing compounds and reinstates the sp^2 structure, which substantially reduces the dispersibility and C/O ratio of the material, making it hydrophobic (Singh et al., 2016). Thermal reduction involves rapid annealing at elevated temperatures ($> 2,000°C$) to induce the exfoliation of GO by the expansion of CO_2, resulting in the decomposition of oxygen functional groups (Stankovich et al., 2007). Thermal reduction has been reported to produce highly graphitic materials with a conductivity of 1,314 S/cm and a C/O ratio of 14.9 (X. Li et al., 2009). With the obvious drawback of high energy consumption and low controllability, chemical methods are widely adapted.

Aqueous dispersions of GO with the reducing agent and a surfactant form graphene nanosheets in a colloidal solution, which is then filtered and dried to obtain rGO (Stankovich et al., 2006). Hydrazine hydrate is the most widely used reducing agent for chemically induced reduction of exfoliated GO. Sodium borohydride has recently proven more efficient than hydrazine in removing C=O species but with leftover alcohol groups (Gao et al., 2009). The principal result of GO reduction is the high current mobility of 320 cm^2/Vs and electrical conductivity of about 6,300 S/cm, close to pristine graphene (Dideikin & Vul', 2019). Even with the severe reduction, a surplus of oxygen functional groups and structural defects will remain due to minimal degradation of carboxylic acids and epoxy groups, and the release of CO_2, respectively. Therefore, the deoxidization of GO does not produce pristine graphene. The partial graphitic surface of rGO makes it adequate for non-covalent functionalization.

6.5 GRAPHENE AND ITS DERIVATIVES IN ENERGY

6.5.1 FUEL CELLS

Fuel cells are energy-converting devices that convert the chemical energy of a fuel into electricity through a redox reaction with eco-friendly by-products (Olabi et al., 2021). The most used fuel in fuel cells is hydrogen. It operates by the electrocatalytic activity of hydrogen oxidation reaction (HOR) and oxidant reduction reactions (ORR) at the electrodes. The electrochemical reaction between hydrogen and oxygen generates water as the by-product, which can be recycled for other applications. Fuel cells are made up of two gas-diffusing layers or electrodes (cathode and anode), an electrolyte, and a catalyst. Chemical reactions occur at the electrodes. The electrolyte transports electrical charges between electrodes and is also known as a proton exchange membrane (PEM). The catalysts accelerate the reactions at the electrodes. The first fuel cell was invented by Sir William Groove in 1838. The main process of fuel cell operation is shown in Figure 6.4. Hydrogen fuel is uniformly channeled through bipolar plates to the anode, while the oxidant is channeled to the cathode. At the anode, catalysis of HOR occurs, causing the hydrogen to split into positive hydrogen ions and negatively charged electrons, as described in equation (6.1). The positive hydrogen ions diffuse through the electrolyte to the cathode, while the electrons travel to an external circuit to the cathode, creating an electrical current. At the cathode, the electrons and positively charged hydrogen ions combine, and catalysis ORR occurs to form water, as given by equation (6.2). The water and heat produced flow out:

$$KMnO_4 + 3H_2SO_4 \rightarrow K^+ + MnO_3^+ + H3O^+ + 3HSO_4^- \qquad (6.1)$$

$$MnO_3^+ + MnO_4^- \rightarrow Mn_2O_7 \qquad (6.2)$$

Classification of fuel cells is based on the component of the electrolyte. They are generally classified into phosphoric acid fuel cells (PAFCs), molten carbonate fuel cells (MCFCs), solid oxide fuel cells (SOFCs), alkaline fuel cells (AFCs), microbial fuel cells (MFCs), and proton exchange membrane fuel cells (PEMFCs). PAFCs, as the first fuel cells to be commercialized, use liquid electrolytes of concentrated phosphoric acid dispersed in PTFE-bonded silicon carbide. MCFCs are high-temperature fuel cells (>600°C) that use molten carbonate salt distributed in a ceramic matrix as an electrolyte. SOFCs utilize solid oxides as electrolytes at 500–700°C to produce electricity and small amounts of carbon dioxide. The most common solid oxide used is yttria-stabilized zirconia (YSZ). AFCs use an alkaline membrane or potassium hydroxide (KOH) adsorbed in a matrix to conduct hydroxide ions. MFCs use microorganisms that are biocatalysts in generating

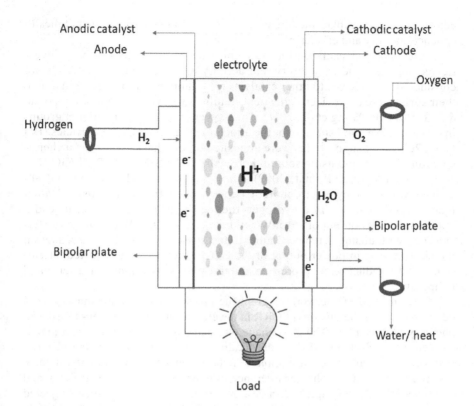

FIGURE 6.4 A classic design of a fuel cell.

electricity via a bioelectrochemical process. The electrolyte material in MFCs is preferably a phosphate buffer because of its almost neutral pH. PEMFCs operate at low temperatures ($< 200°C$) and are the most widely patronized fuel cells because of their low energy/manufacturing cost and the universality of application. They use proton-conducting polymer membrane electrolytes, such as perfluoro sulphonic acid (PFSA), non-fluorinated arylene polymer membranes (e.g., polybenzimidazole, polyether ketone, and poly sulphone), and inorganic proton conductors (e.g., pyrophosphates and solid acids).

With the constant supply of hydrogen and an oxidant, and the removal of byproducts, fuel cells can operate continuously. The aging process of fuel cells, like other energy conversion systems, will limit their performance over time. The rate at which the decay mechanism occurs can be reduced by enhancing the properties of materials present in various components of the fuel cell. Graphene and its derivatives can significantly impact the performance, life span, and durability of fuel cells by increasing ionic and electrical conductivity, electrochemical area, and mechanical and chemical stability (anti-corrosion) properties of its components. Graphene is usually applied as bipolar plates, catalyst support, electrode additives, catalyst replacement, entire electrode, and electrolyte membranes to enhance stability and efficiency.

The support for catalysts on the membrane of fuel cells has been investigated for the issue of stability and activity. Graphene-based materials are considered alternatives to other catalysts (platinum, metal oxides, etc.) due to their surface area and electrochemical stability at elevated temperatures (Guo & Sun, 2012; H. Wang et al., 2011). The oxygen-containing-functional groups in GO provide a good surface for the attachment of catalyst nanoparticles (Bai et al., 2011). Pt nanoparticles were deposited on graphene sheets via a chemical reduction method using sodium borohydride ($NaBH_4$) in H_2PtCl_6/GO (1:1) suspension (Seger & Kamat, 2009). The resulting partially reduced GO-Pt cathode was made by electrophoretic deposition on Toray paper using Nafion ionomer. When tested in a hydrogen fuel cell with Pt: CB anode, it read a maximum power output of 161 mW/cm^3 compared to 96 mW/cm^3 for the reference cell with unsupported platinum. To test thermal and hydrazine reduction methods, a drop in power output of the GO-Pt fuel cell was reported, attributing to the reduction in proton conductivity of Nafion ionomer at elevated temperatures.

The controlled 3D network of graphene provides more active macroporous edges to enhance the kinetics of ORR in the electrocatalyst. Nitrogen-doped graphene synthesized by CVD of methane was demonstrated as an efficient metal-free electrocatalyst for ORR in an alkaline fuel cell via a four-way electron pathway (Qu et al., 2010). It exhibited much better operation stability, notable electrocatalytic activity, higher tolerance to poison (CO in the electrolyte), and crossover effect. A negligible decrease in current was recorded when subjected to −0.1 and 0 V in air-saturated 0.1 M KOH for 200,000 cycles and a steady-state catalytic current thrice of Pt/C.

Although the cathode is mainly responsible for the inactivity of the cell, reducing the amount of Pt in the anode will decrease the presence of platinum in the cell for ample efficiency. A monolithic MFC anode by polyaniline (PANI) hybridized 3D graphene was constructed to facilitate extracellular electron transfer and enhance conductive pathways (Yong et al., 2012). Graphene was synthesized by CVD on nickel substrate using ethanol as a carbon precursor. To ensure proper adhesion of bacteria on the hydrophobic graphene surface, surface modifications with hydrophilic PANI were performed via in situ polymerization. The setup comprised *Shewanella oneidensis* MR-1 and lactate as model electrogene and electron donor, respectively. Electrochemical impedance spectroscopy (EIS) analysis of graphene/PANI MFC, in contrast to standard planar carbon cloth MFC with the same bacterial strain and setup, showed a lower transfer resistance (~100 Ω vs 2,800 Ω), owing to a faster electron transfer rate. The authors conducted an anode performance test in two-chamber MFCs with a loading resistance of 2 kΩ. Graphene/PANI MFC outperformed carbon cloth MFC with 4 times higher maximum power density and 212 times higher specific power density. The latter is due to a lighter graphene/PANI anode than carbon cloth (3 g/m^2 vs 136 g/m^2, respectively).

Graphene has been widely studied as an additive to polymer electrolytes due to its high conductivity performance. High-temperature operations of fuel cells require a highly stable and conductive electrolyte membrane. Phosphonated graphene oxide deposited on 2,6-pyridine functionalized polybenzimidazole (Py-PBI) substrate by solution casting has been investigated (Abouzari-Lotf et al., 2019). Evaluations carried out with Pt-catalyst at high temperatures (120–140°C) under anhydrous conditions recorded a proton conductivity of 76.4×10^{-3} S/cm and optimum power density >359 mW/cm^2. The performance improvement of 70% compared to phosphoric acid–doped Py-PBI membrane was due to the inherently strong hydrogen bonding between the amide group and phosphonic acid group of PGO as well as increased proton diffusion at high temperatures. This result was comparable to conventional Nafion membranes at 80°C under humidified conditions.

In this work, the hole-like self-assembled structure of rolled-up graphene sheets is utilized as a supporting material to fabricate GO/Nafion composite as an electrolyte in PEMCs. Under low humidity conditions, its pristine recast Nafion electrolyte counterpart undergoes proton conductivity loss of 78.14%, while it loses 68.08% due to the rearrangement of the Nafion matrix after surface modification with GO. Tateishi et al. reported the through-plane and in-plane proton conductivities of 10^{-5}–10^{-4} S/cm of multilayered GO film as electrolytes in the absence of Nafion at low RH (<20%) and RT (25°C) (Tateishi et al., 2013). In this work, sulfonic acid functionalized GO paper electrolyte and Pt/C electrode system were used in designing graphene oxide fuel cell (GOFC). The performance was evaluated against the Pt/C electrode and Nafion electrolyte fuel cell. GOFC showed an improved surface area at the Pt/GO interface, which resulted in increased porosity and, consequently, a high power density and efficiency. GO paper is suitable as an electrolyte because of its gas barrier property of effectively separating gases.

Bipolar plates act as current conductors in fuel cells and make up about 80% of the cell weight and 45% of the total cost. It makes them one of the main components that hinder fuel cell commercialization. Copper substrates have been given more attention for practical applications than bipolar plates due to the latter's excellent thermal and electrical conductivities and optical properties. However, it is noteworthy that metals are prone to corrosion when exposed to acidic electrolytes. The interaction between metal bipolar plates and acidic electrolytes can poison the electrocatalyst of the cathode, causing a reduction in power output. Graphene and its derivatives as protective fillers on the surface of bipolar plates can increase corrosion resistance and conductivity without changing their optical properties.

Lee et al. investigated the efficiency of graphene grown through CVD of methane at 830°C as a protective layer on copper bipolar plates (Y. H. Lee et al., 2017). The current-voltage-power (I-V-P) performance of the G/Cu bipolar plate maintained a stable peak power density of 415 mW/cm^2 after 5 h and 235 mW/cm^2 for pure copper bipolar plates. The thin graphene layer coating minimized surface oxidation on copper without performance degradation. The high hydrophobicity of graphene and reduced graphene oxide can increase corrosion resistance by reducing the contact area between the substrate and the corrosive medium. Graphene-octadecyl amine-titanium dioxide (G-ODA-TIO$_2$) nanocomposite was grafted on the surface of Cu using the electrophoretic deposition (ED) method. Drop analysis of the nanocomposites recorded a water contact angle of 130°, 101°, and 87° for G-ODA-TIO$_2$/Cu, G-TIO$_2$/Cu, and pure Cu reference bipolar plates, respectively. The anticorrosion properties were evaluated in 0.5 M H$_2$SO$_4$ electrolyte using potentiodynamic polarization (Tafel analysis). G-ODA-TIO$_2$ recorded the highest corrosion inhibition efficiency of 94% and a rate of corrosion of 0.45 μm/year (Sadeghian et al., 2020).

Improper arrangement of graphene-supporting materials may affect the performance of supported electrocatalysts due to surface interactions and reactivity. The surface area of graphene sheets decreases considerably when separated sheets aggregate together due to the Van der Waals forces and π-π interactions. This influences the activity of fuel cells. Further strategies are required to enhance the conductivity of graphene oxide so the losses that occur in the deposition will not significantly affect the power output of the fuel cell.

6.5.2 Solar Cells

Solar cells are used for the direct harvesting of solar energy. They convert photons to electrons using a semiconductor material to absorb light and generate excitons. Graphene is incorporated in perovskite solar cells (PSCs), Schottky junction solar cells (SJSCs), dye-sensitized solar cells (DSSCs), organic solar cells (OSCs), and so on. It can be utilized as a catalytic counter electrode, transparent conducting cathode, transparent conducting anode, Schottky junction, and active layer materials in solar cells (Patil et al., 2019). These incorporations can improve their stability and quantum efficiency, further improving cell power conversion efficiency (Rehman et al., 2020). The ultrahigh optical transmissivity of graphene creates an optically conducting window for inducing wide ranges of light wavelengths inside solar cells. The high electron mobility of graphene informs a high charge transfer kinetics in solar cells and produces more excellent heat dissipation.

Agresti et al. performed an interface engineering of perovskite solar cells (Agresti et al., 2016). They incorporated graphene flakes (mTiO$_2$ + G) into the mesoporous TiO$_2$ layer and inserted graphene oxide (GO) between the perovskite and hole-transport layers. The process was carried out using a two-step deposition procedure in the air, which resulted in a power conversion efficiency (PCE) of 18.2%. Relative to conventional PSCs, the PCE had increased due to an improved charge-carrier injection/collection. This interface modification also proved favorable for the stability of PSCs under several aging conditions. Hybrid organic-inorganic perovskite as the main component of PSCs has high electron diffusion lengths and great light harvesting. An astonishing PCE of 22.1% has placed them at the forefront of solar cell technology. A change in cell configuration has been actively researched, from the original n-i-p architecture to p-i-n architecture (Nouri et al., 2018). It is to ease the fabrication process, increase stability, and reduce energy requirements while maintaining efficiency. The major drawback in PSCs has been reported as irreversible degradation mainly caused by moisture and metal migration. The hydrophobic nature of graphene prevents humidity and metal atom penetration into the perovskite layer during aging (Domanski et al., 2016).

Conventional Schottky junction solar cells are made of a metal film that tends to absorb most of the solar radiation. ITO as a replacement for metal layers has been reported to overcome this drawback. However, the limited access and brittle nature of ITO increases production cost and obstructs applications in flexible optoelectronic devices, respectively. The merit of near-zero bandgaps and the material universality of graphene make it promising as a graphene/n-type semiconductor Schottky junction. It can partially or fully replace silicon, making graphene-on-silicon Schottky junction or graphene/single nanowire (nanobelt) Schottky junction solar cells, respectively.

To reduce sheet resistance and enhance the built-in potential of graphene-on-silicon SJSCs, p-type doping of graphene with HNO_3 and $SOCl_3$ vapor has been demonstrated with improved PCEs compared to their non-doped counterparts (Fan et al., 2011). The chemical doping of graphene/Si solar cells was investigated by introducing a GO interlayer (Yang et al., 2014). The graphene/GO/Si solar cell was subjected to storage in the open air for a week, and it retained 95% of its original level, indicating improved stability of the solar cell. The advancement in optimization approaches of the graphene/Si Schottky solar cells in recent years has led to an increased PCE from 1.5% to 15.8% in less than a decade (Kong et al., 2019). Key technical challenges remain, such as the graphene wet transfer process on substrates inducing defects, such as tears, cracks, and impurities, such as copper particle residuals (Liang et al., 2011). These defects and impurities can lead to current leakage in the cell and, consequently, affect the efficiency of the device.

A combined electrode approach using 5 nm Au and graphene film in graphene/single NW (NB) SJSCs has been reported. The high work function of Au and graphene transparency and conductivity induced a sheet resistance of 410 Ω per square and transparency larger than 1,200 nm wavelength range, respectively (Ye et al., 2010). Two planar nanostructures of graphene and CdSe nanobelt in Schottky junction solar cell subjected to standard illumination conditions of AM 1.5G and 100 mW/cm³ with similar J-V characteristics on both sides (L. Zhang et al., 2011). Both graphene and CdSe showed an open circuit voltage of 0.49 and 0.48 and PCE of 0.11 and 0.12, respectively. A relatively low fill factor of < 23.7% was reported. It can be attributed to the contact resistance between nanostructures and electrodes, the path of electron transport, or the significant series resistance present in the cell due to the resistance in graphene film.

Graphene is widely utilized in liquid electrolytes of dye-sensitized solar cells. Relative to conventional DSSCs, it prevents the issues of solvent evaporation and temperature instability and ensures easy integration into flexible devices. The oxygen-containing functional groups present in GO and rGO provide active sites for electrocatalysis reactions when applied in counter electrodes of DSSCs. Compared to traditional platinum (Pt)-based counter electrodes, they exhibit astounding performance when doped or hybridized with metals or polymers (Mahmoudi et al., 2018). Graphene extensively mixed with metal/metal oxide nanohybrids as counter electrodes in DSSCs demonstrate excellent electron transfer kinetics (Oh et al., 2020). It is due to the synergistic effect produced by the conductivity of graphene and the interconnectivity between uniformly distributed electrocatalytic metal/metal oxide nanoparticles.

The direct integration of polyimide (PI) on graphene to serve as substrate and carrier film for graphene transfer was investigated for graphene-based electrodes in organic solar cells (Koo et al., 2020). This combination enabled a low sheet resistance of 83 Ω, high optical transmittance of over 92%, and enhanced mechanical durability. The resulting OSC showed inhibited delamination under mechanical stress and a PCE of 15.2%. The charge-transporting layers in OSCs form non-uniform coatings on the surface of graphene due to their hydrophobic and inert nature, resulting in inevitable degradation in device performance. To improve the affinity of graphene surface towards CTLs in OSCs, various alternative routes have been explored:

(a) Solvent-mediated CTLs using polar solvents, such as (PEDOT: PSS) as hole transport layers and ZnO-based solutions as electron transport layers implemented on graphene electrodes both in conventional and inverted configurations (Jung et al., 2017).

(b) Chemical doping of graphene surface (P-type) with $AuCl_3$ in nitromethane to alter the surface wettability of graphene electrodes, improve conductivity, and shift in work function (Park et al., 2010).

Incorporating nanostructures of graphene and its derivatives into photovoltaic devices has produced tailored structures with great flexibility and shown potential for high energy conversion performance. Although a great deal of impressive research has been done in the replacement of conventional materials with graphene-based materials for optoelectronic applications, there is still room for improvement. The sheet resistance of graphene sheets is inversely proportional to the number of layers, according to the following equations:

$$H_2 \rightarrow 2H^+ + 2e^- \tag{6.3}$$

$$2H^+ + 2e^- + 1/2O_2 \rightarrow 2H_2O \tag{6.4}$$

$$R_s = (\sigma_{2D}N) \tag{6.5}$$

where σ_{2D} = bidimensional conductivity of graphene.

The high sheet resistance of graphene results in a high series resistance in solar cells. To reduce the sheet resistance and moderate graphene work function, further research is still required to develop dependable approaches with the following characteristics:

- Controlled sheet size and number of layers
- Bandgap tuning
- Simple fabrication process
- Cost effective
- High yield
- Reproducibility
- Environmentally friendly
- Potential for large-scale production

6.5.3 LITHIUM BATTERIES

Lithium batteries are electrochemical energy storage devices. They range from lithium-ion batteries (LIBs), lithium-air batteries (LABs), lithium-sulfur batteries (LSBs), and so on. Lithium (Li)-ion batteries are widely used in today's mobile electronic devices due to their wider operating voltages and high energy density. The storage capacity of Li-ion batteries is about 150–200 mAh/g, which is still not enough to power electric vehicles. It is required to strike the balance of a lower life span to enable the progress of application in automotive energy storage systems. The efficiency of LIBs is highly dependent on the physicochemical properties of the materials used in the electrodes. The cathode material is restricted to Li, with room to explore the choices of the anode. Typically, the anode is made up of graphite due to its liquid storage capacity and offers a specific increase in storage capacity to about 372 mAh/g with the addition of the Li covalent site (Menachem et al., 1997).

The use of other carbon materials, such as graphene and its derivatives, has been proposed as more suitable anode materials for reversible lithium storage due to their nearly metallic conductance and large surface area. Theoretically, graphene has been reported to exceed graphite with specific capacities of 744 mAh/g and 1,157 mAh/g (Dahn et al., 1995; Sato et al., 1994). Chemical doping and functionalization of graphene can improve its electrochemistry as an anode to further enhance the performance and stability of LIBs. Boron-doped and nitrogen-doped graphene were studied for their high efficiency and long-term cyclability as anodic materials in LIBs (B. Sun et al., 2012). They recorded a peak reversible capacity of >1,040 mAh/g at a low rate of 50 mAh/g with a fast charge and discharge rate within a short period of one hour to several tens of seconds.

Lithium-air batteries (LABs) do not require an intercalation mechanism and have a high charge density, ten times that of lithium-ion batteries (Olabi et al., 2021). Li in the anode reacts directly with O_2 from air, which results in an exponentially increased energy density of 5,200 Wh/kg, making it suitable as a power source for automobiles. Although current LAB systems have been impressive, graphene with an open pore structure is incorporated as a cathode catalyst to increase the round-trip efficiency and stability (Xiao et al., 2011). Graphene was investigated as a high-efficiency cathode catalyst for rechargeable LABs in a non-aqueous electrolyte (M. Yu et al., 2015). The study and analysis of the electrochemical properties of oxygen evolution reaction indicated the occurrence of high catalytic activity.

Lithium-sulfur batteries (LSBs) are currently considered a cheaper alternative to LIBs. They comprise a lithium cathode, a sulfur anode, and a solid or liquid electrolyte. The major drawbacks of LSBs are the issues of volume expansion and contraction during the electrochemical cycling and the electric insulation behavior of sulfur and its discharge product (5×10^{-30} S/cm at 25°C) (M. Yu et al., 2015). Graphene can be used as a scaffold for the sulfur cathode to overcome the mechanical stress and help the redox reaction of all sulfur nanoparticles (M. Yu et al., 2015). The overall performance of lithium batteries is centered on the electrode structure and output, largely reflected by the quality of graphene used.

6.5.4 SUPERCAPACITORS

Supercapacitors are energy storage devices that bridge the gap between rechargeable batteries and conventional capacitors. They are made up of two electrodes, an electrolyte, and a separator. Supercapacitors have a relatively fast charge-discharge rate and serve as a short-term power source. However, they do not store much energy, which hinders their autonomous applications (Castro-Gutiérrez et al., 2020). Increasing the capacitance of supercapacitors highly depends on the composition of the electrode. Different graphene-based electrodes of aerogels, sponges, graphene, and so on are largely used in supercapacitors to improve their energy storage density and efficiency (X. Zhang et al., 2021). Supercapacitors store energy by reversible adsorption of ions at the interfaces between electrolytes and electrode materials. They release energy by the desorption of ions at the interface. Some common types of supercapacitors are pseudo-capacitors (PCs) and electrochemical double-layer capacitors (EDLCs).

Pseudo-capacitors store energy via a faradaic redox reaction. Hydroxides, metal oxides, and electronically conducting polymers potentially improve specific capacitance through pseudo-redox reactions (Lang et al., 2011). This makes them promising electrode materials. However, their application is somewhat limited due to issues of short life span, incompatibility with organic electrolytes, and poor electrical conductivity. Graphene deposited on these pseudo-capacitive materials (e.g., $Ni(OH)_2$) can result in a better electrode-electrolyte interface, increase the conductivity of ions, and consequently, improve the stability and performance of the supercapacitor (Ji et al., 2013).

Electrostatic interactions at the electrode-electrolyte interface in EDLCs lead to the accumulation of charges in the electrical double layer, which requires large surfaces. Thus, graphene-based electrodes are used. For instance, two supercapacitors designed with activated GO produce significantly higher and excellent energy density (approx. 70 Wh/kg) and gravimetric capacitance (150 F/g) with ionic and organic electrolytes (Q. Li et al., 2014).

6.6 CONCLUSION AND FUTURE PERSPECTIVE

This chapter explored the synthesis and current research developments of graphene and its derivatives as next-generation materials for renewable energy technologies. The main and newly developed top-down and bottom-up approaches were reviewed, with a focus on the drawbacks of commercialization. The significant role played by graphene in energy conversion and storage devices, such as fuel cells, solar cells, supercapacitors, and batteries, was discussed. Graphene has the potential to replace conventional materials due to its single-layered 2D structure and unique properties, such as unparalleled thermal conductivity, good adsorption performance, mechanical stability, large surface area, ultra-thinness, and room for functionalization. The feasibility of its applications is largely dependent on its quality, which has been the focus of many studies in the last two decades.

The barriers to the commercialization of graphene were established as low yield, process complexity, high cost, and environmental impact. The ideal graphene is an almost pristine graphene with low oxygen content, few layers, and a wide lateral size. The synthesis of ideal graphene will require control over process parameters to control the morphology of the material to suit intended applications. From a personal perspective, the quality plane is not confined to one method. In this view, method combinations should be further explored to utilize the advantages of different synthesis methods. Research should be geared towards reducing sheet resistance and improving conductivity to solve the issue of stability and power output in energy devices.

The most infant route for graphene synthesis (i.e., flash joule heating) can be the breakthrough for bulk graphene. Although there is limited research on the electronic properties and advanced applications of this method, it has the potential to produce high-quality graphene with low defects. It is the fastest route to graphene production, as it can convert nearly any solid carbon source into valuable graphene flakes in 10 milliseconds. It is also identified as the cheapest and safest route. It utilizes discarded food, rubber wastes, biochar, petroleum waste, and coal, and requires no chemicals or high energy (7.2 kJ of electricity per gram of FLG). This method further reduces greenhouse emissions and carbon footprint. It is catalyst-free and solvent-free, which amplifies its purity. Additionally, the graphene produced by this method is turbostratic graphene, which has misaligned sheets compared to A-B stacked graphene obtained from other exfoliation methods. The layers of turbostratic graphene are easier to pull apart, and so they easily suspend in many solvents. Dedication to fundamental research on FJH could transform the current state of renewable energy technology.

6.7 ACKNOWLEDGMENTS

This work was supported by the University of Brunei Darussalam under research grant number UBD/OAVCRI/CRGWG(022)/171001. Hamidatu Alhassan acknowledges the Ghana Education Trust Fund for financial support.

REFERENCES

Abas, N., Kalair, A., & Khan, N. (2015). Review of fossil fuels and future energy technologies. *Futures*, *69*, 31–49. https://doi.org/10.1016/j.futures.2015.03.003

Abouzari-Lotf, E., Zakeri, M., Nasef, M. M., Miyake, M., Mozarmnia, P., Bazilah, N. A., Emelin, N. F., & Ahmad, A. (2019). Highly durable polybenzimidazole composite membranes with phosphonated graphene oxide for high temperature polymer electrolyte membrane fuel cells. *Journal of Power Sources*, *412*, 238–245. https://doi.org/10.1016/j.jpowsour.2018.11.057

Agresti, A., Pescetelli, S., Taheri, B., Del Rio Castillo, A. E., Cinà, L., Bonaccorso, F., & Di Carlo, A. (2016). Graphene-perovskite solar cells exceed 18% efficiency: A stability study. *ChemSusChem*, *9*(18), 2609–2619. https://doi.org/10.1002/cssc.201600942

Algozeeb, W. A., Savas, P. E., Luong, D. X., Chen, W., Kittrell, C., Bhat, M., Shahsavari, R., & Tour, J. M. (2020). Flash graphene from plastic waste. *ACS Nano*, *14*(11), 15595–15604. https://doi.org/10.1021/acsnano.0c06328

Bai, H., Li, C., & Shi, G. (2011). Functional composite materials based on chemically converted graphene. *Advanced Materials*, *23*(9), 1089–1115. https://doi.org/10.1002/adma.201003753

Banhart, F., Kotakoski, J., & Krasheninnikov, A. V. (2011). Structural defects in graphene. *ACS Nano*, *5*(1), 26–41. https://doi.org/10.1021/nn102598m

Berger, C., Song, Z., Li, T., Li, X., Ogbazghi, A. Y., Feng, R., Dai, Z., Marchenkov, A. N., Conrad, E. H., First, P. N., & de Heer, W. A. (2004). Ultrathin epitaxial graphite: 2D electron gas properties and a route toward graphene-based nanoelectronics. *The Journal of Physical Chemistry B*, *108*(52), 19912–19916. https://doi.org/10.1021/jp040650f

Bhuyan, Md. S. A., Uddin, Md. N., Islam, Md. M., Bipasha, F. A., & Hossain, S. S. (2016). Synthesis of graphene. *International Nano Letters*, *6*(2), 65–83. https://doi.org/10.1007/s40089-015-0176-1

Brodie, B. C. (1860). II. On the atomic weight of graphite. *Proceedings of the Royal Society of London*, *10*(1), 11–12. https://doi.org/10.1098/rspl.1859.0007

Burgess, J. S., Matis, B. R., Robinson, J. T., Bulat, F. A., Keith Perkins, F., Houston, B. H., & Baldwin, J. W. (2011). Tuning the electronic properties of graphene by hydrogenation in a plasma enhanced chemical vapor deposition reactor. *Carbon*, *49*(13), 4420–4426. https://doi.org/10.1016/j.carbon.2011.06.034

Castro-Gutiérrez, J., Celzard, A., & Fierro, V. (2020). Energy storage in supercapacitors: Focus on tannin-derived carbon electrodes. *Frontiers in Materials*, *7*. https://doi.org/10.3389/fmats.2020.00217

Chen, J., Duan, M., & Chen, G. (2012). Continuous mechanical exfoliation of graphene sheets via three-roll mill. *Journal of Materials Chemistry*, *22*(37), 19625. https://doi.org/10.1039/c2jm33740a

Chen, W., Ge, C., Li, J. T., Beckham, J. L., Yuan, Z., Wyss, K. M., Advincula, P. A., Eddy, L., Kittrell, C., Chen, J., Luong, D. X., Carter, R. A., & Tour, J. M. (2022). Heteroatom-Doped Flash Graphene. *ACS Nano*, *16*(4), 6646–6656. https://doi.org/10.1021/acsnano.2c01136

Dahn, J. R., Zheng, T., Liu, Y., & Xue, J. S. (1995). Mechanisms for lithium insertion in carbonaceous materials. *Science*, *270*(5236), 590–593. https://doi.org/10.1126/science.270.5236.590

Dasari, B. L., Nouri, J. M., Brabazon, D., & Naher, S. (2017). Graphene and derivatives—Synthesis techniques, properties and their energy applications. *Energy, 140*, 766–778. https://doi.org/10.1016/j.energy.2017.08.048

Dideikin, A. T., & Vul', A. Y. (2019). Graphene oxide and derivatives: The place in graphene family. *Frontiers in Physics, 6.* https://doi.org/10.3389/fphy.2018.00149

Dimiev, A. M., & Tour, J. M. (2014). Mechanism of graphene oxide formation. *ACS Nano, 8*(3), 3060–3068. https://doi.org/10.1021/nn500606a

Domanski, K., Correa-Baena, J.-P., Mine, N., Nazeeruddin, M. K., Abate, A., Saliba, M., Tress, W., Hagfeldt, A., & Grätzel, M. (2016). Not all that glitters is gold: Metal-migration-induced degradation in perovskite solar cells. *ACS Nano, 10*(6), 6306–6314. https://doi.org/10.1021/acsnano.6b02613

European Commission. (2022, October 14). *Global CO2 Emissions Rebound in 2021 after Temporary Reduction During Covid Lockdown.* https://joint-research-centre.ec.europa.eu/jrc-news/global-co2-emissions-rebound-2021-after-temporary-reduction-during-covid19-lockdown-2022-10-14_en#:~:text=Emissions%20across%20the%20world&text=In%202021%2C%20global%20anthropogenic%20fossil,the%20world%27s%20largest%20CO2%20emitters.

Fan, G., Zhu, H., Wang, K., Wei, J., Li, X., Shu, Q., Guo, N., & Wu, D. (2011). Graphene/silicon nanowire Schottky junction for enhanced light harvesting. *ACS Applied Materials & Interfaces, 3*(3), 721–725. https://doi.org/10.1021/am1010354

Gao, W., Alemany, L. B., Ci, L., & Ajayan, P. M. (2009). New insights into the structure and reduction of graphite oxide. *Nature Chemistry, 1*(5), 403–408. https://doi.org/10.1038/nchem.281

Geim, A. K., & Novoselov, K. S. (2009). The rise of graphene. In *Nanoscience and Technology* (pp. 11–19). Co-Published with Macmillan Publishers Ltd, London, UK. https://doi.org/10.1142/9789814287005_0002

Głuchowski, P., Tomala, R., Jeżowski, A., Szewczyk, D., Macalik, B., Smolina, I., Kurzynowski, T., & Stręk, W. (2020). Preparation and physical characteristics of graphene ceramics. *Scientific Reports, 10*(1), 11121. https://doi.org/10.1038/s41598-020-67977-5

Guo, S., & Sun, S. (2012). FePt nanoparticles assembled on graphene as enhanced catalyst for oxygen reduction reaction. *Journal of the American Chemical Society, 134*(5), 2492–2495. https://doi.org/10.1021/ja2104334

Hayes, C. (2019). True grit [graphene commercial applications]. *Engineering & Technology, 14*(6), 68–71. https://doi.org/10.1049/et.2019.0615

Hegerl, G. C., Brönnimann, S., Cowan, T., Friedman, A. R., Hawkins, E., Iles, C., Müller, W., Schurer, A., & Undorf, S. (2019). Causes of climate change over the historical record. *Environmental Research Letters, 14*(12), 123006. https://doi.org/10.1088/1748-9326/ab4557

Hofrichter, J., Szafranek, B. N., Otto, M., Echtermeyer, T. J., Baus, M., Majerus, A., Geringer, V., Ramsteiner, M., & Kurz, H. (2010). Synthesis of graphene on silicon dioxide by a solid carbon source. *Nano Letters, 10*(1), 36–42. https://doi.org/10.1021/nl902558x

Hummers, W. S., & Offeman, R. E. (1958). *Preparation of Graphitic Oxide.* https://pubs.acs.org/sharingguidelines

International Energy Agency. (2022, March 8). *Global CO2 Emissions Rebounded to Their Highest in History in 2021.* www.iea.org/news/global-co2-emissions-rebounded-to-their-highest-level-in-history-in-2021

Jariwala, D., Srivastava, A., & Ajayan, P. M. (2011). Graphene synthesis and band gap opening. *Journal of Nanoscience and Nanotechnology, 11*(8), 6621–6641. https://doi.org/10.1166/jnn.2011.5001

Ji, J., Zhang, L. L., Ji, H., Li, Y., Zhao, X., Bai, X., Fan, X., Zhang, F., & Ruoff, R. S. (2013). Nanoporous Ni(OH)$_2$ thin film on 3D ultrathin-graphite foam for asymmetric supercapacitor. *ACS Nano, 7*(7), 6237–6243. https://doi.org/10.1021/nn4021955

Jing, N., Xue, Q., Ling, C., Shan, M., Zhang, T., Zhou, X., & Jiao, Z. (2012). Effect of defects on Young's modulus of graphene sheets: a molecular dynamics simulation. *RSC Advances, 2*(24), 9124. https://doi.org/10.1039/c2ra21228e

Jung, S., Lee, J., Choi, Y., Lee, S. M., Yang, C., & Park, H. (2017). Improved interface control for high-performance graphene-based organic solar cells. *2D Materials, 4*(4), 045004. https://doi.org/10.1088/2053-1583/aa823b

Kong, X., Zhang, L., Liu, B., Gao, H., Zhang, Y., Yan, H., & Song, X. (2019). Graphene/Si Schottky solar cells: A review of recent advances and prospects. *RSC Advances, 9*(2), 863–877. https://doi.org/10.1039/C8RA08035F

Koo, D., Jung, S., Seo, J., Jeong, G., Choi, Y., Lee, J., Lee, S. M., Cho, Y., Jeong, M., Lee, J., Oh, J., Yang, C., & Park, H. (2020). Flexible organic solar cells over 15% efficiency with polyimide-integrated graphene electrodes. *Joule, 4*(5), 1021–1034. https://doi.org/10.1016/j.joule.2020.02.012

Lang, X., Hirata, A., Fujita, T., & Chen, M. (2011). Nanoporous metal/oxide hybrid electrodes for electrochemical supercapacitors. *Nature Nanotechnology, 6*(4), 232–236. https://doi.org/10.1038/nnano.2011.13

Lee, X. J., Hiew, B. Y. Z., Lai, K. C., Lee, L. Y., Gan, S., Thangalazhy-Gopakumar, S., & Rigby, S. (2019). Review on graphene and its derivatives: Synthesis methods and potential industrial implementation. *Journal of the Taiwan Institute of Chemical Engineers, 98*, 163–180. https://doi.org/10.1016/j.jtice.2018.10.028

Lee, Y. H., Noh, S., Lee, J.-H., Chun, S.-H., Cha, S. W., & Chang, I. (2017). Durable graphene-coated bipolar plates for polymer electrolyte fuel cells. *International Journal of Hydrogen Energy, 42*(44), 27350–27353. https://doi.org/10.1016/j.ijhydene.2017.09.053

Li, Q., Mahmood, N., Zhu, J., Hou, Y., & Sun, S. (2014). Graphene and its composites with nanoparticles for electrochemical energy applications. *Nano Today, 9*(5), 668–683. https://doi.org/10.1016/j.nantod.2014.09.002

Li, X., Wang, H., Robinson, J. T., Sanchez, H., Diankov, G., & Dai, H. (2009). Simultaneous nitrogen doping and reduction of graphene oxide. *Journal of the American Chemical Society, 131*(43), 15939–15944. https://doi.org/10.1021/ja907098f

Liang, X., Sperling, B. A., Calizo, I., Cheng, G., Hacker, C. A., Zhang, Q., Obeng, Y., Yan, K., Peng, H., Li, Q., Zhu, X., Yuan, H., Hight Walker, A. R., Liu, Z., Peng, L., & Richter, C. A. (2011). Toward clean and crackless transfer of graphene. *ACS Nano, 5*(11), 9144–9153. https://doi.org/10.1021/nn203377t

Lin, T., Chen, J., Bi, H., Wan, D., Huang, F., Xie, X., & Jiang, M. (2013). Facile and economical exfoliation of graphite for mass production of high-quality graphene sheets. *J. Mater. Chem. A, 1*(3), 500–504. https://doi.org/10.1039/C2TA00518B

Lin, T., Tang, Y., Wang, Y., Bi, H., Liu, Z., Huang, F., Xie, X., & Jiang, M. (2013). Scotch-tape-like exfoliation of graphite assisted with elemental sulfur and graphene-sulfur composites for high-performance lithium-sulfur batteries. *Energy & Environmental Science, 6*(4), 1283. https://doi.org/10.1039/c3ee24324a

Liu, F., Wang, C., Sui, X., Riaz, M. A., Xu, M., Wei, L., & Chen, Y. (2019). Synthesis of graphene materials by electrochemical exfoliation: Recent progress and future potential. *Carbon Energy, 1*(2), 173–199. https://doi.org/10.1002/cey2.14

Mahmoudi, T., Wang, Y., & Hahn, Y.-B. (2018). Graphene and its derivatives for solar cells application. *Nano Energy, 47*, 51–65. https://doi.org/10.1016/j.nanoen.2018.02.047

Marcano, D. C., Kosynkin, D. v., Berlin, J. M., Sinitskii, A., Sun, Z., Slesarev, A., Alemany, L. B., Lu, W., & Tour, J. M. (2010). Improved synthesis of graphene oxide. *ACS Nano, 4*(8), 4806–4814. https://doi.org/10.1021/nn1006368

Menachem, C., Peled, E., Burstein, L., & Rosenberg, Y. (1997). Characterization of modified NG7 graphite as an improved anode for lithium-ion batteries. *Journal of Power Sources, 68*(2), 277–282. https://doi.org/10.1016/S0378-7753(96)02629-8

Nouri, E., Mohammadi, M. R., & Lianos, P. (2018). Improving the stability of inverted perovskite solar cells under ambient conditions with graphene-based inorganic charge transporting layers. *Carbon, 126*, 208–214. https://doi.org/10.1016/j.carbon.2017.10.015

Obraztsov, A. N., Obraztsova, E. A., Tyurnina, A. V., & Zolotukhin, A. A. (2007). Chemical vapor deposition of thin graphite films of nanometer thickness. *Carbon, 45*(10), 2017–2021. https://doi.org/10.1016/j.carbon.2007.05.028

Oh, W. C., Cho, K. Y., Jung, C. H., & Areerob, Y. (2020). Hybrid of graphene based on quaternary Cu2ZnNiSe4-WO3 nanorods for counter electrode in dye-sensitized solar cell application. *Scientific Reports, 10*(1), 4738. https://doi.org/10.1038/s41598-020-61363-x

Oh, Y., Nam, S., Wi, S., Kang, J., Hwang, T., Lee, S., Park, H. H., Cabana, J., Kim, C., Park, B., Oh, Y., Nam, S., Wi, S., Kang, J., Hwang, T., Lee, S., Park, P. B., Park, H. H., Cabana, J., & Kim, C. (2013). Supporting Information Effective Wrapping of Graphene on Individual Li 4 Ti 5 O 12 Grains for High-Rate Li-Ion Batteries.

Olabi, A. G., Abdelkareem, M. A., Wilberforce, T., & Sayed, E. T. (2021). Application of graphene in energy storage device—A review. *Renewable and Sustainable Energy Reviews, 135*, 110026. https://doi.org/10.1016/j.rser.2020.110026

Park, H., Rowehl, J. A., Kim, K. K., Bulovic, V., & Kong, J. (2010). Doped graphene electrodes for organic solar cells. *Nanotechnology, 21*(50), 505204. https://doi.org/10.1088/0957-4484/21/50/505204

Parvez, K., Wu, Z.-S., Li, R., Liu, X., Graf, R., Feng, X., & Müllen, K. (2014). Exfoliation of Graphite into Graphene in Aqueous Solutions of Inorganic Salts. *Journal of the American Chemical Society, 136*(16), 6083–6091. https://doi.org/10.1021/ja5017156

Patil, K., Rashidi, S., Wang, H., & Wei, W. (2019). Recent progress of graphene-based photoelectrode materials for dye-sensitized solar cells. *International Journal of Photoenergy, 2019*, 1–16. https://doi.org/10.1155/2019/1812879

Qiu, Y., Hu, Z., Li, H., Ren, Q., Chen, Y., & Hu, S. (2022). Hybrid electrocatalyst Ag/Co/C via flash Joule heating for oxygen reduction reaction in alkaline media. *Chemical Engineering Journal, 430*, 132769. https://doi.org/10.1016/j.cej.2021.132769

Qu, L., Liu, Y., Baek, J.-B., & Dai, L. (2010). Nitrogen-doped graphene as efficient metal-free electrocatalyst for oxygen reduction in fuel cells. *ACS Nano, 4*(3), 1321–1326. https://doi.org/10.1021/nn901850u

Raghavan, N., Thangavel, S., & Venugopal, G. (2017). A short review on preparation of graphene from waste and bioprecursors. *Applied Materials Today, 7*, 246–254. https://doi.org/10.1016/j.apmt.2017.04.005

Rehman, S., Noman, M., Khan, A. D., Saboor, A., Ahmad, M. S., & Khan, H. U. (2020). Synthesis of polyvinyl acetate /graphene nanocomposite and its application as an electrolyte in dye sensitized solar cells. *Optik, 202*, 163591. https://doi.org/10.1016/j.ijleo.2019.163591

Riedl, C., Coletti, C., & Starke, U. (2010). Structural and electronic properties of epitaxial graphene on SiC(0 0 0 1): a review of growth, characterization, transfer doping and hydrogen intercalation. *Journal of Physics D: Applied Physics, 43*(37), 374009. https://doi.org/10.1088/0022-3727/43/37/374009

Sadeghian, Z., Hadidi, M. R., Salehzadeh, D., & Nemati, A. (2020). Hydrophobic octadecylamine-functionalized graphene/TiO2 hybrid coating for corrosion protection of copper bipolar plates in simulated proton exchange membrane fuel cell environment. *International Journal of Hydrogen Energy, 45*(30), 15380–15389. https://doi.org/10.1016/j.ijhydene.2020.04.015

Sadhukhan, S., Ghosh, T. K., Rana, D., Roy, I., Bhattacharyya, A., Sarkar, G., Chakraborty, M., & Chattopadhyay, D. (2016). Studies on synthesis of reduced graphene oxide (RGO) via green route and its electrical property. *Materials Research Bulletin, 79*, 41–51. https://doi.org/10.1016/j.materresbull.2016.02.039

Saeed, M., Alshammari, Y., Majeed, S. A., & Al-Nasrallah, E. (2020). Chemical vapour deposition of graphene—synthesis, characterisation, and applications: A review. *Molecules, 25*(17), 3856. https://doi.org/10.3390/molecules25173856

Salvia, A. L., & Brandli, L. L. (2020). Energy sustainability at universities and its contribution to SDG 7: A systematic literature review. In *Universities as Living Labs for Sustainable Development* (pp. 29–45). Springer. https://doi.org/10.1007/978-3-030-15604-6_3. https://link.springer.com/chapter/10.1007/978-3-030-15604-6_3

Sato, K., Noguchi, M., Demachi, A., Oki, N., & Endo, M. (1994). A mechanism of lithium storage in disordered carbons. *Science, 264*(5158), 556–558. https://doi.org/10.1126/science.264.5158.556

Seger, B., & Kamat, P. v. (2009). Electrocatalytically active graphene-platinum nanocomposites. Role of 2-D carbon support in PEM fuel cells. *The Journal of Physical Chemistry C, 113*(19), 7990–7995. https://doi.org/10.1021/jp900360k

Shams, S. S., Zhang, R., & Zhu, J. (2015). Graphene synthesis: A review. In *Materials Science-Poland* (Vol. 33, Issue 3, pp. 566–578). Walter de Gruyter GmbH. https://doi.org/10.1515/msp-2015-0079

Singh, R. K., Kumar, R., & Singh, D. P. (2016). Graphene oxide: Strategies for synthesis, reduction and frontier applications. *RSC Advances, 6*(69), 64993–65011. https://doi.org/10.1039/C6RA07626B

Somani, P. R., Somani, S. P., & Umeno, M. (2006). Planer nano-graphenes from camphor by CVD. *Chemical Physics Letters, 430*(1–3), 56–59. https://doi.org/10.1016/j.cplett.2006.06.081

Sprinkle, M., Hicks, J., Tejeda, A., Taleb-Ibrahimi, A., le Fèvre, P., Bertran, F., Tinkey, H., Clark, M. C., Soukiassian, P., Martinotti, D., Hass, J., & Conrad, E. H. (2010). Multilayer epitaxial graphene grown on the surface; structure and electronic properties. *Journal of Physics D: Applied Physics, 43*(37), 374006. https://doi.org/10.1088/0022-3727/43/37/374006

Stanford, M. G., Bets, K. V., Luong, D. X., Advincula, P. A., Chen, W., Li, J. T., Wang, Z., McHugh, E. A., Algozeeb, W. A., Yakobson, B. I., & Tour, J. M. (2020). Flash graphene morphologies. *ACS Nano, 14*(10), 13691–13699. https://doi.org/10.1021/acsnano.0c05900

Stankovich, S., Dikin, D. A., Piner, R. D., Kohlhaas, K. A., Kleinhammes, A., Jia, Y., Wu, Y., Nguyen, S. T., & Ruoff, R. S. (2007). Synthesis of graphene-based nanosheets via chemical reduction of exfoliated graphite oxide. *Carbon, 45*(7), 1558–1565. https://doi.org/10.1016/j.carbon.2007.02.034

Stankovich, S., Piner, R. D., Chen, X., Wu, N., Nguyen, S. T., & Ruoff, R. S. (2006). Stable aqueous dispersions of graphitic nanoplatelets via the reduction of exfoliated graphite oxide in the presence of poly(sodium 4-styrenesulfonate). *Journal of Materials Chemistry, 16*(2), 155–158. https://doi.org/10.1039/B512799H

Sun, B., Wang, B., Su, D., Xiao, L., Ahn, H., & Wang, G. (2012). Graphene nanosheets as cathode catalysts for lithium-air batteries with an enhanced electrochemical performance. *Carbon, 50*(2), 727–733. https://doi.org/10.1016/j.carbon.2011.09.040

Sun, L. (2019). Structure and synthesis of graphene oxide. *Chinese Journal of Chemical Engineering, 27*(10), 2251–2260. https://doi.org/10.1016/j.cjche.2019.05.003

Sutter, P. W., Flege, J.-I., & Sutter, E. A. (2008). Epitaxial graphene on ruthenium. *Nature Materials, 7*(5), 406–411. https://doi.org/10.1038/nmat2166

Tarcan, R., Todor-Boer, O., Petrovai, I., Leordean, C., Astilean, S., & Botiz, I. (2020). Reduced graphene oxide today. *Journal of Materials Chemistry C, 8*(4), 1198–1224. https://doi.org/10.1039/C9TC04916A

Tateishi, H., Hatakeyama, K., Ogata, C., Gezuhara, K., Kuroda, J., Funatsu, A., Koinuma, M., Taniguchi, T., Hayami, S., & Matsumoto, Y. (2013). Graphene oxide fuel cell. *Journal of The Electrochemical Society, 160*(11), F1175–F1178. https://doi.org/10.1149/2.008311jes

Tian, Y., Yu, Z., Cao, L., Zhang, X. L., Sun, C., & Wang, D.-W. (2021). Graphene oxide: An emerging electromaterial for energy storage and conversion. *Journal of Energy Chemistry, 55*, 323–344. https://doi.org/10.1016/j.jechem.2020.07.006

Tiwari, S. K., Sahoo, S., Wang, N., & Huczko, A. (2020). Graphene research and their outputs: Status and prospect. In Journal of Science: Advanced Materials and Devices (Vol. 5, Issue 1, pp. 10–29). Elsevier B.V. https://doi.org/10.1016/j.jsamd.2020.01.006

Wang, H., Liang, Y., Li, Y., & Dai, H. (2011). Co1−xS-graphene hybrid: A high-performance metal chalcogenide electrocatalyst for oxygen reduction. *Angewandte Chemie, 123*(46), 11161–11164. https://doi.org/10.1002/ange.201104004

Wang, Y.-C., & Cho, C.-P. (2017). Application of TiO2-graphene nanocomposites to photoanode of dye-sensitized solar cell. *Journal of Photochemistry and Photobiology A: Chemistry, 332*, 1–9. https://doi.org/10.1016/j.jphotochem.2016.07.036

Wet Chemistry of Graphene. (n.d.). https://www.electrochem.org/dl/interface/spr/spr11/spr11_p053-056.pdf

World Economic Forum. (2022, October 10). *Oil Down on Recession Fears, France Fuel Shortages and More: What You Need to Know about the Global Energy Crisis This Week.* www.weforum.org/agenda/2022/10/global-energy-sector-latest-news-10-october/

Xiao, J., Mei, D., Li, X., Xu, W., Wang, D., Graff, G. L., Bennett, W. D., Nie, Z., Saraf, L. V., Aksay, I. A., Liu, J., & Zhang, J.-G. (2011). Hierarchically porous graphene as a lithium—air battery electrode. *Nano Letters, 11*(11), 5071–5078. https://doi.org/10.1021/nl203332e

Xu, C., Luo, G., Liu, Q., Zheng, J., Zhang, Z., Nagase, S., Gao, Z., & Lu, J. (2012). Giant magnetoresistance in silicene nanoribbons. *Nanoscale, 4*(10), 3111. https://doi.org/10.1039/c2nr00037g

Yang, L., Yu, X., Xu, M., Chen, H., & Yang, D. (2014). Interface engineering for efficient and stable chemical-doping-free graphene-on-silicon solar cells by introducing a graphene oxide interlayer. *Journal of Materials Chemistry A, 2*(40), 16877–16883. https://doi.org/10.1039/C4TA02216E

Ye, Y., Dai, Y., Dai, L., Shi, Z., Liu, N., Wang, F., Fu, L., Peng, R., Wen, X., Chen, Z., Liu, Z., & Qin, G. (2010). High-performance single CdS nanowire (Nanobelt) Schottky junction solar cells with Au/graphene Schottky electrodes. *ACS Applied Materials & Interfaces*, *2*(12), 3406–3410. https://doi.org/10.1021/am1007672

Yi, M., & Shen, Z. (2015). A review on mechanical exfoliation for the scalable production of graphene. *Journal of Materials Chemistry A, 3*(22), 11700–11715. https://doi.org/10.1039/C5TA00252D

Yin, J., Molini, A., & Porporato, A. (2020). Impacts of solar intermittency on future photovoltaic reliability. *Nature Communications, 11*(1), 4781. https://doi.org/10.1038/s41467-020-18602-6

Yong, Y.-C., Dong, X.-C., Chan-Park, M. B., Song, H., & Chen, P. (2012). Macroporous and monolithic anode based on polyaniline hybridized three-dimensional graphene for high-performance microbial fuel cells. *ACS Nano, 6*(3), 2394–2400. https://doi.org/10.1021/nn204656d

Yu, M., Li, R., Wu, M., & Shi, G. (2015). Graphene materials for lithium-sulfur batteries. *Energy Storage Materials, 1*, 51–73. https://doi.org/10.1016/j.ensm.2015.08.004

Yu, Q., Lian, J., Siriponglert, S., Li, H., Chen, Y. P., & Pei, S.-S. (2008). Graphene segregated on Ni surfaces and transferred to insulators. *Applied Physics Letters, 93*(11), 113103. https://doi.org/10.1063/1.2982585

Zainal, N., How, J. F., Choo, X. H., & Soon, C. F. (2020). Synthesis and characterization of graphene oxide (GO) and reduced graphene oxide (rGO) using Modified Tour's method for sensing device applications. *2020 IEEE Student Conference on Research and Development (SCOReD)*, 385–390. https://doi.org/10.1109/SCOReD50371.2020.9251005

Zhang, L., Fan, L., Li, Z., Shi, E., Li, X., Li, H., Ji, C., Jia, Y., Wei, J., Wang, K., Zhu, H., Wu, D., & Cao, A. (2011). Graphene-CdSe nanobelt solar cells with tunable configurations. *Nano Research, 4*(9), 891–900. https://doi.org/10.1007/s12274-011-0145-6

Zhang, X., Zhou, J., Zheng, Y., Wei, H., & Su, Z. (2021). Graphene-based hybrid aerogels for energy and environmental applications. *Chemical Engineering Journal, 420*, 129700. https://doi.org/10.1016/j.cej.2021.129700

Zhao, H., & Zhao, T. S. (2013). Graphene sheets fabricated from disposable paper cups as a catalyst support material for fuel cells. *J. Mater. Chem. A, 1*(2), 183–187. https://doi.org/10.1039/C2TA00018K

7 Chitin and Chitosan
Isolation, Deacetylation, and Prospective Biomedical, Cosmetic, and Food Applications

Nurul 'Aqilah Rosman, Mimi Asyiqin Asrahwi,
Nur Alimatul Hakimah Narudin, Mohd Syaadii
Mohd Sahid, Rosmaya Dewi, Norazanita
Shamsuddin, Muhammad Roil Bilad, Eny
Kusrini, Jonathan Hobley, and Anwar Usman

CONTENTS

7.1 INTRODUCTION

Chitosan is one of the natural linear biopolymers. The scientific literature on chitin, which was isolated from a fungus and an exoskeleton of a beetle, was published as early as the 19th century (Khor, 2001; Muzzarelli et al., 2012). A few decades after the first discovery of chitin, the conversion of chitin to chitosan was successfully performed by boiling chitin in a KOH solution (Rouget, 1859; Hoppe-Seyler, 1894). The chemical structures of chitin and chitosan were resolved for the first time using spectroscopic methods in 1950 (Muzzarelli, 1977), and their crystalline properties were reported a decade later (Rudall, 1963). Chitin and chitosan are identified as linear polysaccharides consisting of N-acetylglucosamine and glucosamine moiety, respectively, and have a specific and precise 3D molecular structure, making them much more interesting than random and amorphous structures of other natural polymers or synthetic polymers. It is believed that the semicrystalline structure of chitin and chitosan results from a heterogeneous distribution of acetyl groups, which interlink their polymer strands through hydrogen bonds. The three-dimensional structures of chitin and chitosan are related to their biological functionality and activity (Rameshthangam et al., 2018).

Chitin and chitosan can be enzymatically or chemically modified by introducing different functional groups or complex monomers, allowing the synthesis of a large variety of derivatives with different structures as well as physical, chemical, and biological properties (Elieh-Ali-Komi & Hamblin, 2016). A simple modification, for instance, is shortening the polymer chain, and the different degree of polymerization is the key factor that controls the physical and chemical properties of biopolymers. Over the last few decades, chitin and chitosan, along with other biopolymers, have attracted great attention as promising renewable polymeric materials for their abundancy in nature and their wide range of biomedical, cosmetic, and food applications (Wahab & Razak, 2016; Banachowicz, Gapiński, & Patkowski, 2000).

Chitin and chitosan carry excellent bioactivity properties (Mehrpouya et al., 2021), in particular their non-toxicity, biodegradability, biocompatibility, biomimetic properties, and high availability (Mohiuddin, Kumar, & Haque, 2017; Kumar, 2000; Sinha et al., 2004). Thus, they can be disposed of and quickly degraded by bacteria, and such disposal does not require incineration, which produces carbon dioxide emission, and thus, suppresses spending energy and avoids global environmental issues. This has prompted an important viewpoint where chitin and chitosan, similar to other biopolymers, could be utilized as an alternative to non-biodegradable fossil fuel–based polymers, such as polyethylene terephthalate, polyvinyl chloride, and Styrofoam (Ibrahim et al., 2021; Muthusamy & Pramasivam, 2019; Shaikh, Yaqoob, & Aggarwal, 2021). In addition to the global environmental concern, another important reason is to reduce the use of fossil fuel–based polymers.

The use of these biopolymers as advanced materials for diagnostic, biomedical, cosmetic, food technology, environmental science, biotechnology, material science, and pharmaceutical applications has been intensively explored. It is well known that chitin is insoluble in neutral water due to its semicrystalline structure with intra- and inter-strand hydrogen bonds. Upon deacetylation, the resulting chitosan is also insoluble in neutral water, but it is highly soluble in slightly acidic water due to the protonation of its primary amino groups, so chitosan can be modified easier with other bioactive molecules and is more advantageous for applications as compared to chitin (No & Meyers, 1995). The amino groups of chitosan are important, as this functional group could form intermolecular hydrogen bonding interactions with a wide range of organic compounds, making the conformation of this biopolymer flexible. Therefore, chitosan and its derivatives have good antioxidant (Ivanova & Yaneva, 2020), antimicrobial (Confederat et al., 2021; Yan et al., 2021), hemo-compatible and hemostatic (Naveed et al., 2019), and anticancer activities (Ding & Guo, 2022), and have attracted great interest for a large variety of biomedical applications, such as wound dressing, wound repair, bone tissue engineering, and skin tissue engineering (Madni et al., 2021). Moreover, chitosan is also a potential biomaterial in metal ion scavenging (Varma, Deshpande, & Kennedy, 2004; Al-Rooqi et al., 2022) and drug delivery systems (Parhi, 2020; Bernkop-Schnürch & Dünnhaupt, 2012).

The conventional deacetylation of chitin is performed by heating this biopolymer at 35–105°C in strong acidic or basic solutions. Depending on whether chitin is in a granulate state or completely soluble in the solutions, the deacetylation is categorized as heterogenous or homogenous. Typically, concentrations of the acidic or basic solutions are higher than 5 N (Methacanon et al., 2003; Younes & Rinaudo, 2015). The degree of deacetylation (*DD*) of the resulting chitosan is governed by the concentration of acidic or basic solution as well as the temperature of the medium in the deacetylation reaction (Hahn et al., 2020). However, the polymeric chitosan can only be obtained by deacetylation in basic solutions. This suggests that both heterogeneous or homogenous deacetylation actually destroys or at least shortens the polymeric structures of chitin or chitosan. In recent years, there has been rapid growth in the mechanochemical deacetylation method, which reduces the use of strong acid or base solutions (Chen et al., 2017). More interestingly, this method has been demonstrated to produce chitosan with the degree of polymerization and the molecular weight being adjustable by varying several parameters, including the ratio between the amount of base and chitin, as well as the grinding or milling speed and time (Chen et al., 2017).

In light of the growing interest in the structure-function relationships of chitosan, this chapter provides an overview of the extraction, deacetylation process, deacetylation kinetics of chitin, mechanochemical deacetylation, and the structures and bioactivities of chitosan. This chapter is, therefore, divided into several sections, including structure and characteristics of chitin and chitosan, conventional deacetylation, deacetylation kinetics, mechanochemical deacetylation, antimicrobial activity of chitin and chitosan, and current trends in the chitosan-based materials for biomaterials for diagnostic, biomedical, cosmetics, food technology, environmental sciences, biotechnology, material science, and pharmaceutical applications.

7.2 STRUCTURES AND CHARACTERISTICS OF CHITIN AND CHITOSAN

As described earlier, chitosan is a deacetylated chitin, which is one of the most abundant natural polysaccharides, consisting of long chains of repeating units of monosaccharide monomers linked by glycosidic linkage. Polysaccharides are categorized based on their structure and are commonly found in cells and in cell walls. In this class, chitin is a linear structure of poly(β-(1 $\rightarrow$ 4)-N-acetyl-D-glucosamine), as shown in Figure 7.1, found in the exoskeleton of crustaceans, arachnids, insects, and fungi (Rinaudo, 2006).

Chitin is generally extracted from crustaceans, arachnids, insects, or fungi shells by chemical or biological methods. The chemical method includes a few crucial

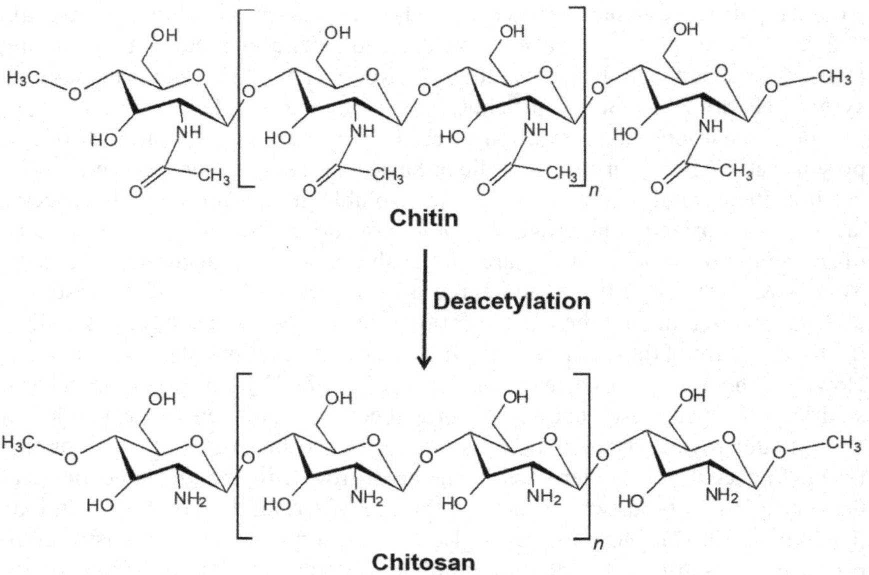

Chitin

Deacetylation

Chitosan

FIGURE 7.1 Deacetylation reaction of chitin to chitosan.

steps, such as demineralization and deproteinization. As the first step, demineralization employs a strong acid, such as HCl, H_2SO_3, H_2SO_4, and HNO_3, to break the calcium carbonate ($CaCO_3$) deposit and dissolve minerals from the shells, and the demineralization process is controlled by the acidity or pH of the employing acid. Although $CaCO_3$ is the most abundant mineral in crustacean shells, the use of a low concentration of strong acids in the demineralization process should be ideal to prevent depolymerization of chitin by the strong acids, as it could shorten the chitin chains due to hydrolysis of the glycosidic bond. The extraction steps include washing, grinding, and sieving the shells. The powdered crustaceans, arachnids, insects, or fungi shells were then demineralized in the solution of a strong acid, such as 1 M HCl, in a 1:40 (w/v) ratio, with stirring for 3 h at room temperature. The foam of carbon dioxide (CO_2) is spontaneously produced when the powdered shells react with the acid, following the reaction given as follows:

$$CaCO_3 \text{ (s)} + 2 \text{ HCl (aq)} \rightarrow CaCl_2 \text{ (aq)} + H_2O \text{ (l)} + CO_2 \text{ (g)} \qquad (7.1)$$

After stirring, the precipitate is filtered, followed by washing with distilled water. The demineralized shells are then dried in an oven at 60–70°C.

The second step, deproteinization, employs strong basic solutions, such as NaOH and KOH with modest concentrations, to remove proteins from the chitin. Here, the dried demineralized shells are deproteinized using 5% NaOH solution in 1:10 (w/v) ratio with stirring for 6 h at 90–95°C, and the mixture is then allowed to cool down, followed by filtration. The obtained precipitate is washed with distilled water to remove any residues. The resulting chitin is dried overnight in an oven at 60–70°C. Decolorization could also be applied as an additional step to remove residual pigments to produce colorless chitin.

The obtained chitin can be deacetylated using a strong acidic or basic solution across different parameters, such as the concentration of acidic or basic solutions, temperature, and deacetylation time, according to the reaction scheme presented in Figure 7.1 (Narudin et al., 2022; Narudin et al., 2020). The obtained deacetylated chitin precipitate, which is chitosan, was isolated by vacuum filtration, followed by washing the precipitate several times with distilled water until it achieved pH 7. The obtained chitosan was dried overnight in an oven at 60–70°C. The deacetylation kinetics was monitored by varying the stirring duration for different deacetylation times.

7.3 CHARACTERIZATIONS OF CHITIN AND CHITOSAN

The most important characteristic of chitosan is its *DD* value, as it determines other physical and chemical properties of this biopolymer. Although the *DD* value of chitosan may depend on both the source and the isolation method of chitin, the deacetylation of chitin plays a crucial role in the physical and chemical properties of the resulting chitosan. Firstly, the deacetylation of chitin is represented by the deacetylation yield, which is analyzed by comparing the weight of chitin before and after the deacetylation reaction. Secondly, deacetylation is reflected by the modification of the functional groups and chemical structures of chitin, which are observable in vibrational spectra, such as Fourier-transform infrared (FTIR) or Raman spectra. FTIR spectroscopy is also considered a simple technique to determine the *DD* value of chitosan.

Figure 7.2 shows the FTIR spectra of chitin and chitosan. Typically, chitin shows a few broad and intense vibrational bands. The vibrational band of chitin

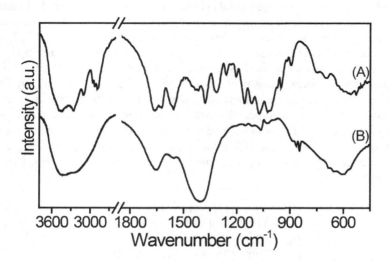

FIGURE 7.2 FTIR spectra of (A) chitin and (B) mud chitosan.

appeared at around 3,450 cm^{-1} due to the formation of its intramolecular hydrogen-bonded OH group. The vibrational stretching modes of amide I appear as two intense bands at 3,264 cm^{-1} and 3,110 cm^{-1} due to its intermolecular hydrogen bonding interactions with amide carbonyl (C=O) group and its intramolecular hydrogen bonding interactions (Marchessault, Pearson, & Liang, 1960). In the fingerprint region, the bands are at 1,662 and 1,630 cm^{-1}, which are assigned to doublet vibrational peaks of amide I of chitin interconnected by intramolecular and intermolecular hydrogen bonding interactions (Focher et al., 1992; Kumari et al., 2015). The spectral pattern of amide I reflects the intramolecular and intermolecular hydrogen bonds, which stabilize the three-dimensional structure of chitin. In addition, the FTIR spectrum of chitin also shows bands at 2,861–2,971 cm^{-1}, 1,424 cm^{-1}, and 1,380 cm^{-1}, which are assigned to the vibrational modes for CH stretching, CH$_2$ bending, and CH$_3$ distortion, respectively.

Deacetylation of chitin leads to abrupt changes in the FTIR spectrum, suggesting that the acetyl group of chitins is removed upon deacetylation, followed by the formation of NH$_2$ of chitosan (see Figure 7.1). Typically, the FTIR spectrum of chitosan has a broad band at 2,850–3,650 cm^{-1} with two peaks at 3,445 cm^{-1} and 3,221 cm^{-1} due to OH and NH$_2$ stretching vibrations. In the fingerprint region, chitosan shows spectral bands at 1,655 cm^{-1} of the carbonyl of amide group, at 1,600 cm^{-1} due to deformation of NH$_2$ group, and at 1,406 cm^{-1} due to CH group. In particular, the vibrations of pyranose group of chitosan appear as a broad band in the range of 484–825 cm^{-1}. It is also interesting to highlight that (i) the doublet vibrational peaks of amide I of chitin related to its α-form disappear and (ii) the two intense bands at 3,264 and 3,110 cm^{-1} due to the NH of the amide group of chitins disappear and turn into a small shoulder band. These facts suggested that hydrogen binding interactions involving the amide group also disappear in chitosan.

Chitosan shows a single band at 1,655 cm^{-1}, suggesting that this biopolymer is interconnected by hydrogen bonds (Focher et al., 1990). The amide I band of chitosan is also identified as a well-defined band at 1,467 cm^{-1}. The presence of N-acetyl-glucosamine, which is a distinct band for chitosan appears at 1,322 cm^{-1}. The vibrational band observed at 1,159 cm^{-1} is attributed to an asymmetrical bridge with oxygen stretching as well as asymmetrical in-phase ring stretching vibrational modes. The peak intensity at around 1,655 cm^{-1} and that of C=O stretching for the secondary amide are influenced by how much the DD value of chitosan is.

Based on the vibrational peak intensity of the amide band of the acetyl group and hydroxyl group of chitosan at 1,655 cm^{-1} and 3,450 cm^{-1}, as denoted respectively by A_{1655} and A_{3450}, the DD value is calculated using the following equations (Baxter et al., 1992; Küçükgülmez, 2018; Sabnis & Block, 1997; Domszy & Roberts, 1985):

$$DD = 100 - [115 \times A_{1655}/A_{3450}] \tag{7.2}$$
$$DD = 118.883 - [10.1647 \times A_{1655}/A_{3450}] \tag{7.3}$$
$$DD = 97.67 - [26.486 \times A_{1655}/A_{3450}] \tag{7.4}$$
$$DD = 100 - [75.188 \times (A_{1655}/A_{3450})] \tag{7.5}$$

It is interesting to note that chitosan is naturally hygroscopic and chitosan with lower *DD* values may absorb more moisture as compared to those with higher *DD* values. Thus, to determine its *DD* value, it is essential that the chitosan under analysis should be completely dry (Blair et al., 1987).

The other important characteristics of chitin and chitosan are the degree of acetylation (DA) and molecular weight. The DA value of isolated chitin and chitosan can be determined by the spectroscopic, enzymatic, and chromatographic methods (Baxter et al., 1992; Küçükgülmez, 2018; Sabnis & Block, 1997; Domszy & Roberts, 1985). The DA represents the fraction of N-acetyl-glucosamine units in chitin and chitosan chains. Therefore, DA 50% is considered as the borderline between chitin and chitosan. The DA value of chitin and chitosan is an essential parameter that determines the physical and chemical properties of these polysaccharides.

The distribution of molecular weight of chitin depends on its source and the demineralization process. The use of highly concentrated HCl in the demineralization reduces the degree of polymerization of chitin. The molecular weight of chitin is determined using the chromatographic method or is quantified based on its intrinsic viscosity using the established Mark-Houwink equation (Brugnerotto et al., 2001):

$$[\eta] = KM^a \tag{7.6}$$

Here, $[\eta]$ is the intrinsic viscosity, and K and a are constants. The values of K and a vary between $5.7 \times 10^{-2} - 8.2 \times 10^{-2}$ and $0.76-0.83$, depending on the DA of chitosan and experimental conditions. The high values of a suggest that chitin and chitosan are semi-rigid structures of polysaccharides. The measured molecular weight is an average value of a large distribution of the molecular weight of chitin and chitosan chains.

The crystallinity of chitin and chitosan was analyzed using the X-ray diffraction (XRD) method. Based on its polymer arrangement, chitin crystallizes into three different polymorphs; the α-chitin has an antiparallel chain arrangement, the β-chitin has a parallel chain arrangement, and the γ- chitin is a mixture of α- and β-chitin, as illustrated in Figure 7.3. In these chitin polymorphs, the different chain arrangements result in distinct crystal structures and intermolecular interactions. Although, in all the crystalline polymorphs, chitin chains are interconnected by intra-sheet C=O···H–N, C–OH···O=C, and C–OH···OH–C hydrogen bonds, the sheets of α-chitin are further linked by additional C–OH···OH–C hydrogen bonding interactions. In particular, the absence of inter-sheet interactions allows the penetration of small solvent molecules, such as water and alcohol, into the β-chitin structure, making the β-chitin swell rapidly in these kinds of solutions. Therefore, XRD helps in determining the atomic and molecular structures of chitin and the level of purity. The latter is presented using a crystallinity index.

Figure 7.4 shows typical XRD patterns of α-chitin and α-chitosan within the diffraction 2θ angles of $20-60°$. The main diffraction peak of α-chitin is observed

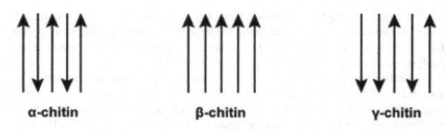

FIGURE 7.3   The structure of chitin with three different orientations of polymer strands, as distinguished to be α-, β-, and γ-polymorphs.

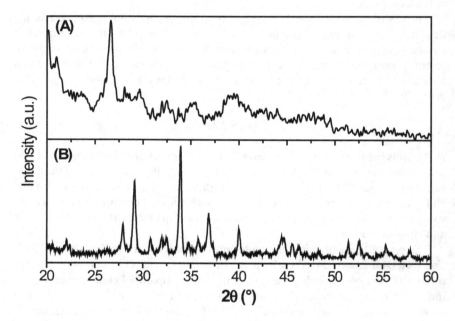

FIGURE 7.4   XRD patterns of (A) chitin and (B) chitosan.

at $2\theta = 26.5°$ due to its (130) plane, and other minor peaks appear at higher diffraction angles (Cárdenas et al., 2004). It is noteworthy that, similar to chitin, chitosan also has a linear structure and crystallizes into three different polymorphs (i.e., α-, β-, and γ-chitosan). The main diffraction peaks of α-chitosan are observed at 27.9°, 29.2°, 33.9°, 36.9°, 40.0°, 44.6°, 45.6°, 46.3°, 51.5°, 52.6°, 55.3°, and 57.8°. Although the crystal structure of both α-chitin and α-chitosan belongs to the orthorhombic P212121 space symmetry group (Cárdenas et al., 2004), the removal of the acetyl group of chitin upon deacetylation leads to slightly larger d spacing of the crystalline structure of chitosan as compared to that of chitin. In particular, the crystal lattice parallel to the inter-strand of chitosan is 0.895 nm, longer than that of chitin (0.469 nm) (Cárdenas et al., 2004). This means that as compared to chitin, chitosan strands are arranged in a less compact manner in its three-dimensional crystal structure. Both chitin and chitosan exhibit partially an amorphous structure; the ratio of the crystalline portion and the amorphous portion is presented as the crystallinity index, which is defined as the relative area of diffraction peaks with respect to the total area under the curve (Abdou et al., 2008). The crystallinity index of chitin and chitosan extracted from crustacean shells is within 60–75% (Mohan et al., 2019).

Chitin extracted from the crustacean shells, insect cuticles, and fungal cell walls mostly belongs to the α-polymorph. However, chitin extracted from squid feathers, squid pens, the tubes of tubeworms, and chaetae of annelids is the β-chitin, and that extracted from spider beetle, locust, cockroach, mantis, dragonfly, silkworm larva is the γ-chitin (Daraghmeh et al., 2011; Arbia et al., 2013; Kaya et al., 2017). As summarized in Table 7.1, the chitin content varies in the range of 2–75%, depending on the chitin source (Synowiecki & Al-Khateeb, 2003). As crustacean shells give a high yield of chitin, the sources for commercial chitin are mainly crab, shrimp, and lobster shells.

As gathered in Table 7.2, the chitin extracted from crustacean shells has been pointed out to have residual ash and moisture contents less than 2.1% and 9.0%, respectively, and possess a degree of acetylation (DA) in the range of 53.0–88.5%. The intrinsic viscosity, $[\eta]$, of chitin obtained from crustacean shells is

TABLE 7.1
The Quantum Yield of Chitin Isolation from Different Organisms

Organism	Chitin (%)	Organism	Chitin (%)
Crustaceans:		**Insects:**	
Cancer (crab)	72.1	*Periplaneta* (cockroach)	2.0
Carcinus (crab)	64.2	*Blatella* (cockroach)	18.4
Paralithodes (king crab)	35.0	*Coleopetra* (ladybird)	27–35
Callinectes (blue crab)	14.0	*Diptera* (house fly)	54.8
Crangon and *Pandalus* (shrimp)	17–40	*Pieris* (butterfly)	64.0
Alaska shrimp	28.0	**Fungi:**	
Nephron (lobster)	69.8	*Bombyx* (silkworm)	44.2
Homarus (lobster)	60–75	*Galleria* (wax worm)	33.7
Lepas (goose barnacle)	58.3	*Aspergillus niger*	42.0
Scylla serata (mud crab)	17.0	*Penicillium chryogenum*	20.1
Mollusks:		*Penicillium chryogenum*	20.1
Clam	6.1	*Saccharomyces cerevisiae*	2.9
Shell oyster	3.6	*Mucor rouxii*	44.5
Squid pen	41.0	*Lactarius vellereus*	19.0
Krill	40.2		

TABLE 7.2

Physicochemical Properties of Chitin and Chitosan

	Mud Crab	Crab	Shrimp
Chitin			
Residual ash	2.1%	0.5	0.4
Moisture	9.083%		
%DA	73.2–76.4%	78.6%	88.5%
Chitosan			
Yield	84.7%	53.0%	74.5%
Residual ash	2.0%		
Moisture	8.47%		
$[\eta]$	93 mL g^{-1}	78 mL g^{-1}	175 mL g^{-1}
M	8.96 kDa	6.12 kDa	17.03 kDa
%DA	11.5%	17%	12%

between 78 and 175 mL g^{-1}, from which its viscosity-average molecular weight (M) has been estimated to be in the range of 8.96–17.03 kDa. It is worth noting that the DA of chitosan derived from crustacean shells is around 11–17%, indicating incomplete deacetylation of chitin.

7.4　DEACETYLATION OF CHITIN

Chitin is converted into chitosan via deacetylation using alkali or enzymatic procedures. Through the deacetylation process, chitosan is obtained by removing the acetyl group (CH$_3$-CO) of chitin, leaving behind an amino group (–NH$_2$). As the deacetylation of chitin is almost never complete (Tavares et al., 2020), and the final product still contains N-acetyl-D-glucosamine (acetylated) units, which are distributed randomly with β-(1-4)-linked D-glucosamine (deacetylated unit) throughout the structure of chitosan (No & Meyers, 1995). In this regard, the *DD* value is used to represent the molar fraction of D-glucosamine in the chitosan. Typically, chitosan can be divided into high (85–95%), middle (70–85%), and low (55–70%) *DD* values, whereas chitosan with an ultrahigh *DD* value (95–100%) can only be achieved by several consecutive deacetylation processes (Alvarenga et al., 2010). Regardless of its *DD* value, chitosan has therefore hydroxyl and amino groups which are active and capable to form intermolecular hydrogen bonds with a variety of chemicals or to undergo different chemical reactions. The most important reactions include esterification, etherification, carboxylation, and hydroxylation, which introduce new pendant groups, improving biological activities and physicochemical properties, and enhancing the solubility of the chemically modified chitosan.

　　As mentioned earlier, heterogeneous deacetylation is mostly performed by mixing the solid particles of chitin with acidic or basic solutions (Chen et al.,

2017). This chemical process is much more efficient compared with the enzymatic process (Galed et al., 2005). In the basic solution, the hydroxyl (OH^-) ions react with acetamide groups, producing solid chitosan and acetate ion, which is dissolved in the solution and finally forms sodium acetate, as given by the following (Ahlafi et al., 2013):

$$R - NHCOCH_3 + NaOH \rightarrow R - NH_3 + CH_3COONa \qquad (7.7)$$

where R represents the chitin polymer backbone. This reaction produces another OH^- ion, which deprotonates an intermediate species to generate a dianion, which converts into chitosan. As a prolonged reaction time is often required to achieve a high conversion from chitin to chitosan and the biopolymers can undergo hydrolysis and depolymerization in a strong basic solution, the use of a modest concentration of an alkaline solution is preferred in the deacetylation process. With the chemical process, one can adjust a few parameters, such as the concentration of reagent, temperature, and reaction time to achieve an optimized condition.

Several studies have explored the deacetylation of chitin, exploring the effects of the concentration of reagent, temperature, and reaction time on the deacetylation kinetics (Kasaai, 2009; Tolaimate et al., 2000). The kinetics and mechanisms of deacetylation of chitin have been described based on the observed *DD* value across different parameters. The deacetylation is considered a noncatalytic liquid-solid reaction and should randomly take place in a homogeneous alkaline solution (de Souza & Giudici, 2021). The deacetylation kinetics is evaluated based on the shrinking core model (de Souza & Giudici, 2021), where the external layer is considered to be deacetylated first, followed by acetyl groups inside the chitin particle. With the shrinking core model, under non-isothermal conditions, the deacetylation converting the acetyl group to the amino group is a pseudo-first-order reaction, and the kinetics can be expressed as follows (Kurita, Sannan, & Iwakura, 1977):

$$d[R - NHCOCH_3]/dt = -k[R - NHCOCH_3][NaOH] \qquad (7.8)$$

The remaining concentration of chitin is represented by the DA value of the resulting chitosan. Here, k is the deacetylation rate. For an excessive amount of NaOH in the solution, [NaOH] is relatively unchanged along the reaction, and hence, equation (7.8) can be written as follows:

$$d[R - NHCOCH_3]/dt = -k[R - NHCOCH_3] \qquad (7.9)$$

Considering the amount of the acetyl group in the chitin before and after deacetylation at time t, equation (7.9) straightforwardly gives the following:

$$ln\{([R - NHCOCH_3]_0 - [R - NHCOCH_3]_t)/[R - NHCOCH_3]_0\} = -kt \qquad (7.10)$$

or

$$-ln\{1 - DD\} = kt \qquad (7.11)$$

Here, DD denotes the ratio of $[R - NHCOCH_3]_t$ with respect to $[R - NHCOCH_3]_0$.

The DD value of the resulting chitosan increases nonlinearly with the deacetylation time. Typically, deacetylation takes place in two stages; that is, a faster stage at early reaction time and a slower stage at longer reaction times (Narudin et al., 2022). To clarify this finding, Figure 7.5 shows the plot of $-ln\{1 - DD\}$ against t at different temperatures, demonstrating two linear regions separated by one transition point. This indicates that the deacetylation of chitin occurs in two different stages, and each stage follows the pseudo-first-order. The rate of the earlier stage is about one order of magnitude higher than the later stage (Narudin et al., 2022). The faster rate is due to the deacetylation of acetamide groups of the amorphous region on the external layer of the chitin particle, whereas the slower rate is due to those of the crystalline region inside the chitin particle (Kurita, Sannan, & Iwakura, 1977). This suggests that, in the heterogeneous deacetylation, (i) acetamide groups of the amorphous region on the external layer of the chitin particle are deacetylated easier than those of the crystalline region inside the chitin particle and (ii) the rate-determining step is the diffusion of OH⁻ ions onto the surface and then into the crystalline region inside the chitin particle (Sarhan et al., 2009). The transition point is shifted to an earlier time at a higher temperature, as acetamide groups in the crystalline region inside the chitin particle are attacked easier at a higher temperature due to structural changes, larger swelling of chitin particles, or higher dynamics of OH⁻ ions at higher temperatures. The concentration of NaOH is another important factor that controls the deacetylation kinetics. In this regard, the deacetylation rate increases with the quadratic concentration of NaOH (Khong, Aachmann, & Vrum, 2012).

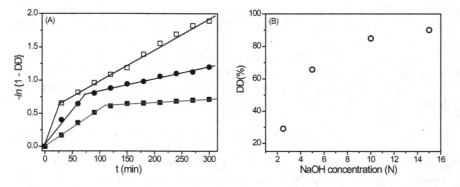

FIGURE 7.5 (A) Plots of $-ln\{1 - DD\}$ as a function of reaction time, t, of the deacetylation reaction using 10 N NaOH at 35°C (■), 75°C (●), and 105°C (□); and (B) DD as a function of the NaOH concentration.

The deacetylation kinetics of chitin is governed by the activation energy (E_a) of the reaction between the acetamide groups of chitins and OH⁻ ions. In this sense, the deacetylation is performed at different temperatures. Based on the temperature-dependent deacetylation rate constant, the activation energy (E_a) of the deacetylation is determined using the Arrhenius equation:

$$k = A \, exp \, (-E_d/RT) \tag{7.12}$$

Here, A is the pre-exponential factor, R is the gas constant (8.314 J mol⁻¹ K⁻¹), and T is temperature. It has been reported in many studies that the deacetylation of acetamide groups in the amorphous region on the external layer of chitin particles has lower activation energy as compared with those in the crystalline region inside the chitin particle (Lavertu, Darras, & Buschmann, 2012). Typically, the activation energy of the deacetylation of chitin derived from crustacean shells is in the range of 16.2–56.0 kJ mol⁻¹ (Ahlafi et al., 2013; Yaghobi & Hormozi, 2010). The wide distribution of the energy activation of the heterogeneous deacetylation might be related to different physicochemical properties, such as the DA, degree of polymerization, molecular weight, crystallinity index, solubility, and particle size distribution of chitin, which govern the structural changes and swelling factor of chitin particle as well as the diffusion and penetration of OH⁻ ions into the chitin particle and, thus, the activation energy.

For homogeneous deacetylation, the reaction between the acetamide groups of chitins and OH⁻ ions should be solely controlled by the diffusion of the chitin strands and OH⁻ ions in the mixture. Interestingly, the activation energy of homogeneous deacetylation of chitin extracted from crustacean shells is 92 kJ mol⁻¹ (Sannan, Kurita, & Iwakura, 1977), which is much higher than those of heterogeneous deacetylation. The plausible reason for the higher activation energy of homogeneous deacetylation of chitin could be the dynamics of diffusions of both the chitin strands and OH⁻ ions.

Recently, alternative deacetylation of chitin used the mechanochemical method. This method includes chemical synthesis that is triggered by external mechanical work, such as when two solids are ground together in a shaker, mortar and pestle, or ball mill (Solares-Briones et al., 2021). An extension to this technique is liquid-assisted grinding, where a minimal amount of liquid is employed as an addition to increase and/or control reactivity in a solvent-free mechanochemical method (Ying, Yu, & Su, 2021). There are only a few studies on the mechanochemical methods in the synthesis of chitin to chitosan, and the majority of it is using the ball milling technique, which finely grounds and mixes bulky materials into nanosized particles with various ball sizes. The size and material of the balls, the mixing time and speed, and the size of the container used can possibly alter the size of the nanostructure (de Arquer et al., 2021). A study reported by Nardo et al. demonstrated a mechanochemical method to synthesize chitosan using the ball milling method (Nardo et al., 2019), where grinding the mixture of chitin and 5% NaOH at 1:5 (w/v) ratio with the addition of a small amount of solvent for 30–90 minutes results in 23% DD of chitosan

(Nardo et al., 2019). In another study, Anusha et al. used chitin extracted from squid pens to synthesize chitosan via the ball mill method (Anusha et al., 2016). The ratio used was 1:15 (w/v) of chitin powder to 50% NaOH in a nitrogen gas chamber and stirred for 2 hours. It was found that the DD for this chitosan sample was about 80% (Anusha et al., 2016).

7.5 ANTIMICROBIAL ACTIVITY OF CHITIN AND CHITOSAN

The antimicrobial activity of chitin and chitosan has attracted great interest. These biopolymers with a concentration as high as 1% have been shown to inhibit the growth of *Staphylococcus aureus* after incubation in slightly acidic conditions (pH 5.5–6.5) for two days [72]. In a few antibacterial studies, chitosan with lower concentrations has been demonstrated to inhibit completely the growth of several bacterial strains, including *S. aureus*, *Bacillus cereus*, *Escherichia coli*, and *Proteus vulgaris* (Chang et al., 1989; Darmadji & Izumimoto, 1994; Simpson et al., 1997). In all cases, chitin and chitosan are suggested to have bactericidal effects. However, numerous studies showed the inactivity of chitin and chitosan against bacterial strains. The reasons are the neutral pH of medium and low concentrations of the biopolymers used in their experiment. The large variation of the minimum concentration of chitin and chitosan to inhibit the growth of bacterial strains reflects the mechanism of action of the biopolymers, as it has been pointed out to depend on the DA of the biopolymers. The role of the pH of the medium is also crucial, where the primary amino groups of chitosan are protonated at pH 5.5–6.5, and this protonated amine is responsible for the antibacterial activity. With the same argument, the solution of chitosan has been reported to be no longer bactericidal at pH 7, at which the amino groups are not protonated (Lim et al., 2021).

7.6 APPLICATIONS OF CHITOSAN

Due to its distinctive quality, including cationic character, film-forming properties, and biological activity, chitosan has a wide range of applications in different biomedical and manufacturing industries, such as food, cosmetics, and pharmaceuticals, as summarized in Table 7.3. For such applications, chitosan, with a top-grade purity and high DD value, is required. In particular, positively charged chitosan mixed with negatively charged heparin produces a stable heparin-chitosan complex that encourages re-epithelialization of full-thickness wounds in human skin (Emanuelsson & Kratz, 1997). Recently, chitosan was applied in biomedical applications, such as colonic, gastrointestinal, nasal, and ophthalmic applications. In addition, as mentioned earlier, chitosan has become a favored biocompatible and biodegradable material for drug carriers to deliver drugs onto the target organs and to prevent drug irritation (Miyazaki, Ishii, & Nadai, 1981; Humbatova et al., 2016). The bottom line is that with its rich hydroxyl and amino groups, chitosan can adsorb organic pharmaceutical compounds efficiently. Intermolecular interactions of chitosan with other organic or inorganic materials

TABLE 7.3

The Applications of Chitosan

Biomedical	Tissue engineering
	Burn treatment
	Contact lenses
	Wound dressing
	Wound repair
	Bone tissue engineering
	Skin tissue engineering
	Drug carriers
Cosmetics	Bath lotions
	Hair conditioner
	Hair care products
	Moisturizer
	Face and body cream
Foods	Food coatings
	Edible films
	Whipping, emulsifying, gelling, stabilizing, and thickening agents

have prompted the formation of chitosan composites, which can act as scaffolds for tissue engineering (Kim et al., 2008; Chow & Khor, 2000). In this sense, the structures of chitosan composites form strong porous scaffolds. For example, chitosan-hydroxyapatite composites have been used as biomaterials in the repair and reconstruction of bone (Kokubo, 1991; Huang et al., 2005; Sarasam & Madihally, 2005; Hsieh et al., 2005; Chung et al., 2002). It is worth noting that chitosan composites also have good properties in terms of structural strength and oxygen permeability, in addition to their biocompatibility. Therefore, chitosan-based biomaterials are applicable for wound dressings and burn treatment (Azuma et al., 2015). In this sense, the efficacy of chitosan composites–based biomaterials to induce collagen synthesis in wound dressings and burn treatment has been investigated (Miyazaki, Ishii, & Nadai, 1981; Bano et al., 2017). As it has been reported, chitosan-based biomaterials accelerate wound contraction (Nowroozi et al., 2021) due to their analgesic and hemostatic properties.

It is important to note that chitosan has to be in nanometer scales. The chitin and chitosan nanoparticles can be prepared by various top-down methods, and most of them are ionotropic gelation, microemulsion, and microwave methods, employing sodium tripolyphosphate, glutaraldehyde, and potassium persulfate as reagents. The sodium tripolyphosphate forms chitosan strands into particles, glutaraldehyde makes microemulsion of chitosan, while potassium persulfate cuts chitosan strands into shorter polymeric chains. The formation of the chitosan nanoparticles occurs rapidly. The size distributions of the chitosan particles are within a few tens of nanometers to a few hundred nanometers. The small size of chitosan nanoparticles provides an excellent feature for drug delivery and imaging. The chitosan-based nanoparticles can transport the drugs to target organs to treat diseases.

In cosmetic industries, chitin and chitosan have been used in skin, nail, oral, hair conditioner, and hair care products. The use of chitin and chitosan in cosmetics is due mainly to their cleansing, UV light–protecting, humectant, and antioxidant properties. Chitosan is the primary ingredient, as it is physiologically safe and is free from any harmful monomers due to an incomplete polymerization process. The use of chitosan also facilitates interactions with proteins so that they are less statically charged during brushing and combing and more stable at high humidity than other conventional hair-treated fixatives (Semwal, Singh, & Dutta, 2013).

Polysaccharides, including chitosan, pectin, and carrageenan, are widely used in food processing industries to give beneficial qualities like whipping, emulsifying, gelling, stabilizing, and thickening effects to food products. Chitosan, without exception, has been shown to have some distinguishing properties (Knorr, 1984). It was found that chitosan has significantly higher water uptake than microcrystalline cellulose (Knorr, 1982). The efficiency of chitosan for retrieval of suspended particles and coagulation in processing wastes from poultry, eggs, seafood, and crop industries has also been shown in several studies (Venugopal, 2021; Nouj et al., 2021). These studies showed that the suspended particles of multiple food manufacturing wastes can be minimized using chitosan.

The ability of chitosan to form films and its biodegradable nature have prompted the use of this biopolymer as coatings and edible films to replace polyethylene films (Kester & Fennema, 1986; Labuza & Breene, 1989). The latter is known to have some disadvantages, as the use of polyethylene films accelerates the condensation of water and depletes penetration of oxygen, promoting fermentation and fungal growth (El Ghaouth et al., 1992; Miller, Spalding, & Risse, 1983). The use of chitosan also improves the quality and extends the shelf life of fresh, frozen, and fabricated foods. In this sense, coating the chitosan on the outer layer of foods can control the morphological and physicochemical properties of the food products. In particular, chitosan films could control moisture and oxygen penetration into the foods, thus inhibiting fermentation and the growth of microbes, and decreasing the rate of metabolism.

7.7   CONCLUSION

In summary, numerous research efforts have been intensively devoted to the extraction of chitin, deacetylation of chitin to chitosan, and development of chitin and chitosan applications. Chitin can be extracted from the exoskeleton of crustaceans, arachnids, insects, and fungi. Chitin and chitosan are very promising advanced materials for biomedical, cosmetic, and food applications. While both chitin and chitosan possess antioxidant, antimicrobial, and anticancer activities, they are biocompatible, biodegradable, non-toxic, and non-immunogenic. The physical and chemical properties and bioactivities of chitosan depend on the source and isolation method of chitin, especially the deacetylation process. Chitosan can also be chemically modified to introduce new pendant groups, improving the biological activities and physicochemical properties, and enhancing the solubility of the biopolymer. Chitin and chitosan have been demonstrated to have a wide range of biomedical applications in wound dressing, wound repair, bone tissue engineering, and skin tissue engineering. In addition, chitin and chitosan nanoparticles can be prepared in nanometer scales using various top-down methods, making chitin and chitosan excellent materials in drug delivery systems. Due to their cleansing, UV light–protecting, humectant, and antioxidant properties, chitin and chitosan have been applied in various cosmetic products. Chitin and chitosan give beneficial qualities like whipping, emulsifying, gelling, stabilizing, and thickening effects to food products. In addition, the use of chitosan also improves the quality and extends the shelf life of fresh, frozen, and fabricated foods. In addition to the established deacetylation of chitin by heating it in acid or basic solution, recent findings on mechanochemical deacetylation to produce chitosan are of interest. Further development of mechanochemical deacetylation of chitin to chitosan in the future would be able to decipher the detailed physicochemical properties of the resulting chitosan. This would open new improvements and explorations of chitosan as an advanced material for further diagnostic, biomedical, cosmetic, food, and pharmaceutical applications.

REFERENCES

Abdou ES, Nagy KSA, Elsabee MZ. 2008. Extraction and characterization of chitin and chitosan from local sources. *Bioresour. Technol.* 99(5): 1359–1367.

Ahlafi H, Moussout H, Boukhlifi F, Echetna M, Bennani MN, Slimane MS. 2013. Kinetics of N-deacetylation of chitin extracted from shrimp shells collected from coastal area of Morocco. *Mediterr. J. Chem.* 2(3): 503–513.

Al-Rooqi MM, Hassan MM, Moussa Z, Obaid RJ, Suman NH, Wagner MH, Natto SSA, Ahmed SA. 2022. Advancement of chitin and chitosan as promising biomaterials. *J. Saudi Chem. Soc.* 26: 101561.

Alvarenga ES, de Oliveira CP, Bellato CR. 2010. An approach to understanding the deacetylation degree of chitosan. *Carbohydr. Polym.* 80(4): 1155–1160.

Anusha JR, Fleming AT, Arasu MV, Kim BC, Al-Dhabi NA, Yu K-H, Raj CJ. 2016. Mechanochemical synthesis of chitosan submicron particles from the gladius of *Todarodes pacificus. J. Adv. Res.* 7(6): 863–871.

Arbia W, Arbia L, Adour L, Amrane A. 2013. Chitin extraction from crustacean shells using biological methods. *Food Technol. Biotechnol.* 51(1): 12–25.

Azuma K, Izumi R, Osaki T, Ifuku S, Morimoto M, Saimoto H. 2015. Chitin, chitosan, and its derivatives for wound healing: Old and new materials. *J. Funct. Biomater.* 6: 104–142.

Banachowicz E, Gapiński J, Patkowski A. 2000. Solution structure of biopolymers: A new method of constructing a bead model. *Biophys. J.* 78(1): 70–78.

Bano I, Arshad M, Yasin T, Ghauri MA, Younus M. 2017. Chitosan: A potential biopolymer for wound management. *Int. J. Biol. Macromol.* 102: 380–383.

Baxter A, Dillon M, Taylor KD, Roberts GAF. 1992. Improved method for IR determination of the degree of N-acetylation of chitosan. *Int. J. Biol. Macromol.* 14: 166–169.

Bernkop-Schnürch A, Dünnhaupt S. 2012. Chitosan-based drug delivery systems. *Eur. J. Pharm. Biopharm.* 81(3): 463–469.

Blair HS, Guthrie J, Law T-K, Turkington P. 1987. Chitosan and modified chitosan membranes I. Preparation and characterisation. *J. Appl. Polym. Sci.* 33(2): 641–656.

Brugnerotto J, Desbrières J, Roberts G, Rinaudo M. 2001. Characterization of chitosan by steric exclusion chromatography. *Polymer* 42(25): 9921–9927.

Cárdenas G, Cabrera G, Taboada E, Miranda SP. 2004. Chitin characterization by SEM, FTIR, XRD, and 13C cross polarization/mass angle spinning NMR. *J. Appl. Polym. Sci.* 93(4): 1876–1885.

Chang DS, Cho HR, Goo HY, Choe WK. 1989. A development of food preservation with the waste of crab processing. *Bull. Korean Fish Soc.* 22: 70–78.

Chen X, Yang H, Zhong Z, Yan N. 2017. Base-catalysed, one-step mechanochemical conversion of chitin and shrimp shells into low molecular weight chitosan. *Green Chem.* 19: 2783.

Chow KS, Khor E. 2000. Novel fabrication of open-pore chitin matrixes. *Biomacromolecules* 1: 61–67.

Chung TW, Yang J, Akaike T, Cho KY, Nah JW, Kim SI. 2002. Preparation of alginate/galactosylated chitosan scaffold for hepatocyte attachment. *Biomaterials* 23: 2827–2834.

Confederat LG, Tuchilus CG, Dragan M, Sha'at M, Dragostin OM. 2021. Preparation and antimicrobial activity of chitosan and its derivatives: A concise review. *Molecules* 26: 3694.

Daraghmeh NH, Chowdhry BZ, Leharne SA, Al Omari MM, Badwan AA. 2011. Chitin. In: *Profiles of Drug Substances, Excipients and Related Methodology*, Academic Press Inc., London, UK, Vol. 36, pp. 35–102. https://www.sciencedirect.com/bookseries/profiles-of-drug-substances-excipients-and-related-methodology

Darmadji P, Izumimoto M. 1994. Effect of chitosan in meat preservation. *Meat Sci.* 38: 243–254.

de Arquer FPG, Talapin DV, Klimov VI, Arakawa Y, Bayer M, Sargent EH. 2021. Semiconductor quantum dots: Technological progress and future challenges. *Science* 373(6555): eaaz8541.

de Souza JR, Giudici R. 2021. Effect of diffusional limitations on the kinetics of deacetylation of chitin/chitosan. *Carbohydr. Polym.* 254: 117278.

Ding J, Guo Y. 2022. Recent advances in chitosan and its derivatives in cancer treatment. *Front. Pharmacol.* 13: 888740.

Domszy JG, Roberts GAF. 1985. Evaluation of infrared spectroscopic techniques for analysing chitosan. *Makromol. Chem.* 186(8): 1671–1677.

El Ghaouth A, Ponnampalam R, Castaigne F, Arul J. 1992. Chitosan coating to extend the storage life of tomatoes. *Hortscience* 27: 1016–1018.

Elieh-Ali-Komi D, Hamblin MR. 2016. Chitin and chitosan: Production and application of versatile biomedical nanomaterials. *Int. J. Adv. Res.* 4(3): 411–427.

Emanuelsson P, Kratz G. 1997. Characterization of a new in vitro burn wound model. *Burns* 23(1): 32–36.

Focher B, Beltrame PL, Naggi A, Torri G. 1990. Alkaline N-deacetylation of chitin enhanced by flash treatments—reaction kinetics and structure modifications. *Carbohydr. Polym.* 405: 18–31.

Focher B, Naggi A, Torri G, Cosani A, Terbojevich M. 1992. Structural differences between chitin polymorphs and their precipitates from solutions—evidence from CP-MAS 13C-NMR, FT-IR and FT-Raman spectroscopy. *Carbohydr. Polym.* 17(2): 97–102.

Galed G, Miralles B, Panos I, Santiago A, Heras A. 2005. N-Deacetylation and depolymerization reactions of chitin/chitosan: Influence of the source of chitin. *Carbohydr. Polym.* 62(4): 316–320.

Hahn T, Tafi E, Paul A, Salvia R, Falabella P, Zibek S. 2020. Current state of chitin purification and chitosan production from insects. *J. Chem. Technol. Biotechnol.* 95(11): 2775–2795.

Hoppe-Seyler F. 1894. Ueber chitosan und zellulose. *Ber. Deut. Chem.: Gesell.* 27: 3329–3331.

Hsieh C-Y, Tsai S-P, Wang D-M, Chang Y-N, Hsieh H-J. 2005. Preparation of γ-PGA/chitosan composite tissue engineering matrices. *Biomaterials* 26: 5617–5623.

Huang Y, Onyeri S, Siewe M, Moshfeghian A, Madihally SV. 2005. In vitro characterization of chitosangelatin scaffolds for tissue engineering. *Biomaterials* 26: 7616–7627.

Humbatova SF, Zeynalov NA, Taghiyev DB, Tapdiqov SZ, Mammedova SM. 2016. Chitosan polymer composite material containing of silver nanoparticle. *Dig. J. Nanomater. Biostruct.* 11(1): 39–44.

Ibrahim NI, Shahar FS, Sultan MTH, Md Shah AU, Safri SNA, Mat Yazik MH. 2021. Overview of bioplastic introduction and its applications in product packaging. *Coatings* 11: 1423.

Ivanova DG, Yaneva ZL. 2020. Antioxidant properties and redox-modulating activity of chitosan and its derivatives: Biomaterials with application in cancer therapy. *Biores. Open Access* 9(1): 64–72.

Kasaai MR. 2009. Various methods for determination of the degree of N-acetylation of chitin and chitosan: A review. *J. Agric. Food Chem.* 57(5): 1667–1676.

Kaya M, Mujtaba M, Ehrlich H, Salaberria AM, Baran T, Amemiya CT, Galli R, Akyuz L, Sargin I, Labidi J. 2017. On chemistry of γ-chitin. *Carbohydr. Polym.* 176, 177–186.

Kester JJ, Fennema OR. 1986. Edible films and coatings; A review. *Food Technol.* 40: 47–59.

Khong TT, Aachmann FL, Vrum KM. 2012. Kinetics of de-N-acetylation of the chitin disaccharide in aqueous sodium hydroxide solution. *Carbohydr. Res.* 352: 82–87.

Khor E. 2001, *Chitin: Fulfilling a Biomaterials Promise*, Elsevier Science Limited, Oxford.

Kim I-Y, Seo S-J, Moon H-S, Yoo M-K, Park I-Y, Kim A-C. 2008. Chitosan and its derivatives for tissue engineering applications. *Biotechnol. Adv.* 26: 1–21.

Knorr D. 1982. Functional properties of chitin and chitosan. *J. Food Sci.* 47(2): 593–595.

Knorr D. 1984. Use of chitinous polymers in food – a challenge for food research and development. *Food Technol.* 38: 85–97.

Kokubo T. 1991. Bioactive glass ceramics: Properties and applications. *Biomaterials* 12: 155–163.

Küçükgülmez A. 2018. Extraction of chitin from crayfish (*Astacus leptodactylus*) shell waste. *Alınteri J. Agric. Sci.* 33(1): 99–104.

Kumar MNVR. 2000. A review of chitin and chitosan applications. *React. Funct. Polym.* 46(1): 1–27.

Kumari S, Rath P, Kumar ASH, Tiwari TN. 2015. Extraction and characterization of chitin and chitosan from fishery waste by chemical method. *Environ. Technol. Innov.* 3: 77–85.

Kurita K, Sannan, T, Iwakura Y. 1977. Studies on chitin; Evidence for formation of block and random copolymers of N-acetyl-D-glucosamine and D-glucosamine by hetero- and homogeneous hydrolyses. *Makromol. Chem.* 178(12): 3197–3202.

Labuza TP, Breene WM. 1989. Applications of active packaging for improvement of shelf-life and nutritional quality of fresh and extended shelf-life foods. *J. Food Proc. Preserv.* 13: 1–69

Lavertu M, Darras V, Buschmann MD. 2012. Kinetics and efficiency of chitosan reacetylation. *Carbohydr. Polym.* 87: 1192–1198.

Lim MJ, Shahri NNM, Taha H, Mahadi AH, Kusrini E, Lim J-W, Usman A. 2021. Biocompatible chitin-encapsulated CdS quantum dots: Fabrication and antibacterial screening. *Carbohydr. Polym.* 260: 117806.

Madni A, Kousar R, Naeema N, Wahid F. 2021. Recent advancements in applications of chitosan-based biomaterials for skin tissue engineering. *J. Bioresour. Bioprod.* 6(1): 11–25.

Marchessault RH, Pearson FG, Liang CY. 1960. Infrared spectra of crystalline polysaccharides. *Biochim. Biophys. Acta* 45: 499–507.

Mehrpouya M, Vahabi H, Barletta M, Laheurte P, Langlois V. 2021. Additive manufacturing of polyhydroxyalkanoates (PHAs) biopolymers: Materials, printing techniques, and applications. *Mater. Sci. Eng. C* 127: 112216.

Methacanon P, Prasitsilp M, Pothsree T, Pattaraarchachai J. 2003. Heterogeneous *N*-deacetylation of squid chitin in alkaline solution. *Carbohydr. Polym.* 52(2): 119–123.

Miller WR, Spalding DH, Risse LA. 1983. Decay, firmness and color development of Florida bell pepper dipped in chlorine and imazalil and film-wrapped. *Proc. Fla. State Hort. Soc.* 96: 347–350.

Miyazaki S, Ishii K, Nadai T. 1981. The use of chitin and chitosan as drug carriers. *Chem. Pharm. Bull.* 29: 3067–3069.

Mohan K, Ravichandran S, Muralisankar T, Uthayakumar V, Chandirasekar R, Rajeevgandhi C, Rajan DK, Palaniappan Seedevi P. 2019. Extraction and characterization of chitin from sea snail *Conus inscriptus* (Reeve, 1843). *Int. J. Biol. Macromol.* 126: 555–560.

Mohiuddin M, Kumar B, Haque S. 2017. Biopolymer composites in photovoltaics and photodetectors. In: *Biopolymer Composites in Electronics*; Sadasivuni KK, Ponnamma D, Kim J, Cabibihan JJ, AlMaadeed MA. (Eds.), Elsevier Inc., Amsterdam, pp. 459–486.

Muthusamy MS, Pramasivam S. 2019. Bioplastics—An eco-friendly alternative to petrochemical plastics. *Curr. World Environ.* 14: 49–59.

Muzzarelli RAA. 1977. *Chitin*, Pergamon Press Ltd: Hungary, pp. 83–143.

Muzzarelli RAA, Boudrant J, Meyer D, Manno N, DeMarchis M, Paoletti MG. 2012. Current views on fungal chitin/chitosan, human chitinases, food preservation, glucans, pectins and inulin: A tribute to Henri Braconnot, precursor of the carbohydrate polymers science, on the chitin bicentennial. *Carbohydr. Polym.* 87: 995–1012.

Nardo TD, Hadad C, Nhien ANV, Moores A. 2019. Synthesis of high molecular weight chitosan from chitin by mechanochemistry and aging. *Green Chem.* 21: 3276–3285.

Narudin NAH, Mahadi AH, Kusrini E, Usman A. 2020. Chitin, chitosan, and submicronsized chitosan particles prepared from *Scylla serrata* shells. *Mater. Int.* 2: 139–149.

Narudin NAH, Rosman AR, Shahrin EWES, Sofyan N, Mahadi AH, Kusrini E, Hobley J, Usman A. 2022. Extraction, characterization, and kinetics of N-deacetylation of chitin obtained from mud crab shells. *Polym. Polym. Compos.* 30: 09673911221109611.

Naveed M, Phil L, Sohail M, Hasnat M, Baig MMFA, Ihsan AU, Shumzaid M, Kakar MU, Akabar TMKMD, Hussain MI, Zhou Q-G. 2019. Chitosan oligosaccharide (COS): An overview. *Int. J. Biol. Macromol.* 129: 827–843.

No HK, Meyers SP. 1995. Preparation and characterization of chitin and chitosan—A review. *J. Aqua. Food Product Technol.* 4: 27–51.

Nouj N, Hafid N, El Alem N, Cretescu I. 2021. Novel liquid chitosan-based biocoagulant for treatment optimization of fish processing wastewater from a Moroccan plant. *Materials* 14: 7133.

Nowroozi N, Faraji S, Nouralishahi A, Shahrousvand M. 2021. Biological and structural properties of graphene oxide/curcumin nanocomposite incorporated chitosan as a scaffold for wound healing application. *Life Sci.* 264: 118640.

Parhi R. 2020. Drug delivery applications of chitin and chitosan: A review. *Environ. Chem. Lett.* 18: 577–594.

Rameshthangam P, Solairaj D, Arunachalam G, Ramasamy P. 2018. Chitin and chitinases: Biomedical and environmental applications of chitin and its derivatives. *J. Enzyme* 1: 20–43.

Rinaudo M. 2006. Chitin and chitosan: Properties and applications. *Prog. Polym. Sci.* 31(7): 603–632.

Rouget C. 1859. Des substances amylacées dans les tissus des animaux, spécialement des articulés (chitine). *Comp. Rend* 48: 792–795.

Rudall KM. 1963. The chitin/protein complexes of insect cuticles. *Adv. Insect Physiol.* 257–314.

Sabnis S, Block LH. 1997. Improved infrared spectroscopic method for the analysis of degree of N-deacetylation of chitosan. *Polym. Bull.* 39: 67–71.

Sannan T, Kurita K, Iwakura Y. 1977. Studies on chitin. V. Kinetics of deacetylation reaction. *Polym. J.* 9(6): 649–651.

Sarasam A, Madihally SV. 2005. Characterization of chitosan polycaprolactone blends for tissue engineering applications. *Biomaterials* 26: 5500–5508.

Sarhan AA, Ayad DM, Badawy DS, Monier M. 2009. Phase transfer catalyzed heterogeneous N-deacetylation of chitin in alkaline solution. *React. Funct. Polym.* 69(6): 358–363.

Semwal A, Singh R, Dutta PK. 2013. Chitosan: A promising substrate for pharmaceuticals. *J. Chitin Chitosan Sci.* 1: 1–16.

Shaikh S, Yaqoob M, Aggarwal P. 2021. An overview of biodegradable packaging in food industry. *Curr. Res. Food Sci.* 4: 503–520.

Sinha VR, Singla AK, Wadhawan S, Kaushik R, Kumria R, Bansal K, Dhawan S. 2004. Chitosan microspheres as a potential carrier for drugs. *Int. J. Pharm.* 274(1–2): 1–33.

Simpson BK, Gagne N, Ashie INA, Noroozi E. 1997. Utilization of chitosan for preservation of raw shrimp (*Pandalus borealis*). *Food Biotechnol.* 11: 25–44.

Solares-Briones M, Coyote-Dotor G, Páez-Franco JC, Zermeño-Ortega MR, de la O Contreras CM, Canseco-González D, Avila-Sorrosa A, Morales-Morales D, Germán-Acacio JM. 2021. Mechanochemistry: A green approach in the preparation of pharmaceutical cocrystals. *Pharmaceutics* 13(6): 790.

Synowiecki J, Al-Khateeb NA. 2003. Production, properties, and some new applications of chitin and its derivatives. *Crit. Rev. Food Sci. Nutr.* 43(2): 145–171.

Tavares L, Flores EEE, Rodrigues RC, Hertz PF, Noreña CPZ. 2020. Effect of deacetylation degree of chitosan on rheological properties and physical chemical characteristics of genipin-crosslinked chitosan beads. *Food Hydrocoll.* 106: 105876.

Tolaimate A, Desbrières J, Rhazi M, Alagui A, Vincendon M, Vottero P. 2000. On the influence of deacetylation process on the physicochemical characteristics of chitosan from squid chitin. *Polymer* 41(7): 2463–2469.

Varma AJ, Deshpande SV, Kennedy JF. 2004. Metal complexation by chitosan and its derivatives: A review. *Carbohydr. Polym.* 55: 77–93.

Venugopal V. 2021. Valorization of seafood processing discards: bioconversion and bio-refinery approaches. *Front. Sustain. Food Syst.* 5: 611835.

Wahab IF, Razak SIA. 2016. Polysaccharides as composite biomaterials. In: *Composites from Renewable and Sustainable Materials*; Poletto M. (Ed.), IntechOpen, Rijeka, Croatia, pp. 65–84. https://www.intechopen.com/books/5440

Yaghobi N, Hormozi F. 2010. Multistage deacetylation of chitin: Kinetics study. *Carbohydr. Polym.* 81(4): 892–896.

Yan D, Li Y, Liu Y, Li N, Zhang X, Yan C. 2021. Antimicrobial properties of chitosan and chitosan derivatives in the treatment of enteric infections. *Molecules* 26: 7136.

Ying P, Yu J, Su W. 2021. Liquid-assisted grinding mechanochemistry in the synthesis of pharmaceuticals. *Adv. Synth. Catal.* 363(5): 1246–1271.

Younes I, Rinaudo M. 2015. Chitin and chitosan preparation from marine sources. structure, properties and applications. *Mar. Drugs* 13: 1133–1174.

8 Quantum Dots
From Synthesis to Biomedical and Biological Applications

*Nurulizzatul Ningsheh Mohammad Shahri,
Taqiyyuddin Akram Aidil, Junaidi H. Samat,
Cristina Pei Ying Kong, Nurul Amanina A.
Suhaimi, Ensan Waatriah E. S. Shahrin, Aisyah
Farhanah Abdul Majid, Abdul Hanif Mahadi,
Muhammad Nur, and Anwar Usman*

CONTENTS

8.1 INTRODUCTION

The quantum dot (QD), which is referred to as a semiconductor nanocrystal, is one of the forefronts of nanotechnology. QDs were discovered for the first time by Alexey Ekimov in 1981 during doping a semiconductor in glass (Ekimov & Onushchenko, 1981). After the discovery, the absorption and emission properties of QDs that are linked to their particle size have been realized (Ekimov, Efros, & Onushchenko, 1985). Research into this nanomaterial has opened up new

DOI: 10.1201/9781003367819-8

horizons in physics, chemistry, and biology. Advanced science and technology have prompted further the use of QDs in a wide area, involving interdisciplinary collaborations.

The most exciting property of QDs is their small sizes, between 2 and 10 nm in diameter, shorter than the bulk exciton Bohr radius, facilitating a strong quantum confinement effect (Valizadeh, Mikaeili, & Samiei, 2012). Moreover, a high surface-area-to-volume ratio of QDs has attracted great attention and has led to a wide range of applications (Alivisatos, 1996). In particular, quantum confinement results in an intense variety of energy and can be observed in the electronic and fluorescent spectra different from those in bulk (Apter et al., 2018). Additionally, quantum confinement can also be observed with the strength of Coulombic interactions, provided that the diameter of the material is the same magnitude as the de Broglie wavelength of the electron wave function. Excitons are formed as the energy of hollow-electron increases, thus, allowing radiative recombination and a narrow band of phonon emission (Pinaud et al., 2006). The confined exciton in QDs has shown unique optical and optoelectronic properties, photoluminescence, and laser effects (Apter et al., 2018). Due to the quantum size effect, QDs exhibit a narrow size distribution, as shown in Figure 8.1. Therefore, the colloidal QDs are considered a new kind of fluorophore, where the particles exhibit different narrow emission bands, depending on their sizes. However, the fluorescence quantum yield is strongly determined by the exciton-hole separation; thus,

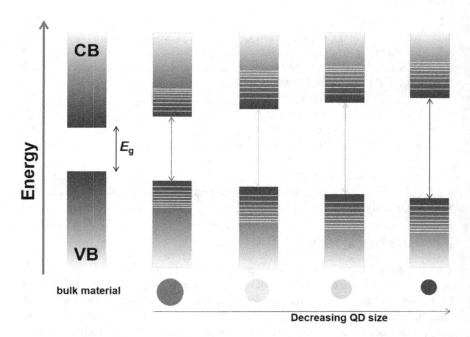

FIGURE 8.1 The bandgap (E_g) increases with the decrease in semiconductor QD size. CB and VB denote the conduction and valence bands of the bulk semiconductor structure.

suppressing the exciton-hole recombination has become an important approach to increase the fluorescence quantum yield of QDs.

Capping agents that are used to passivate the metallic atoms on the particle surface in order to stabilize the formation of QDs could control the physical and chemical properties of QDs. Therefore, proper design and selection of a suitable capping agent, for instance, could be used to improve the solubility of QDs in aqueous buffers, enhance the exciton lifetime, prevent QDs from agglomeration, or stop uncontrolled growth. In this regard, small thiol compounds, such as thioglycolic acid (TGA), mercaptopropionic acid (MPA), and mercaptoundecanoic acid (MUA), are conventionally and commonly used capping agents. The TGA-, MPA-, and MUA-capped metal chalcogenide QDs are well dispersed in water. Numerous research efforts have been devoted to developing new QDs with well-designed capping agents to enhance their functionality.

Strong π-conjugated capping agents used to stabilize the formation of QDs could facilitate the electron transfer surrounding molecules and play an essential role in interactions with target analytes (Shahri et al., 2022; Suarez et al., 2017). QDs capped with strong biological affinity or biofunctional molecules have been used as biological probes, diagnostic, and therapeutic agents (Purushothaman & Song, 2021; Blasiak, van Veggel, & Tomanek, 2013; McMillan, Batrakova, & Gendelman, 2011; Pisanic, Zhang, & Wang, 2014). For example, CdTe QDs conjugated with the tumor-specific ligands and PbS QD capped with micelle of diagnostic agent have been used for imaging of tumors in vivo (Li et al., 2012) and cancer cells (Hu et al., 2012), respectively. Due to their distinctive photostability, distinctive design and engineering of capping agents, and rich surface chemistry, QDs can be used as traceable drug nanocarriers in drug delivery systems. In this sense, QDs have been conjugated with captopril (Manabe et al., 2006). The pharmacokinetics and pharmacodynamics of the antihypertensive agent have been successfully evaluated in stroke-prone hypertensive rats (Manabe et al., 2006). In addition, QDs conjugated with doxorubicin, a DNA-interacting drug, have been utilized in chemotherapy (Baqalkot et al., 2007). Driven by well-established applications of QDs in the field of medical science, this chapter is focused on the applications of QDs in biological probes, disease diagnostics, therapeutics, and drug delivery systems. This chapter comprises the fabrication and conjugation of QDs, surface properties of QDs, and applications of QDs in biological probes, disease diagnostics, therapeutics, and drug delivery systems.

8.2   FABRICATION AND CONJUGATION OF QDS

In general, two main approaches have been established to synthesize nanomaterials and QDs; namely, top-down and bottom-up. The top-down synthesis relies on direct cleaving or breaking down bulk materials into nanoscale particles via physical processes, such as mechanical milling (Protesescu et al., 2018), photolithography (Antolini & Orazi, 2019), ultrasonic-assisted or microwave-assisted process (Kandasamy et al., 2021), and laser ablation of bulky materials to small-sized particles (Horoz et al., 2012), or via chemical processes, such as electrochemical oxidation (Liu et al., 2016), solvothermal (Takezawa et al., 2021), and hydrothermal (Yang & Park, 2019).

In the top-down methods, mechanical processing, for instance, requires high temperatures and/or pressures to obtain a metastable crystalline phase, which is stable at room temperature and ambient pressure (Protesescu et al., 2018). In contrast, photolithography employs photoresists and optical lithography (Antolini & Orazi, 2019), while laser ablation utilizes high fluence pulsed laser (Biswas et al., 2012). The advantage of these top-down methods is that their synthetic process is low cost, allowing the synthesis to be suitable for the production of high quantities of QDs, while the disadvantages of these methods are the non-selectivity of any chemical reactions and extreme synthetic conditions, leading to poor control over the size and morphology of QDs.

In contrast, bottom-up approaches are based on the gradual growth of small organic precursor molecules through QD chemical reactions of precursors or molecular building blocks (de Arquer et al., 2021). The formation of QDs is governed by the Stranski-Krastanov growth model (de Arquer et al., 2021). This approach plays an important role in the fabrication and processing of nanostructures and nanomaterials, and they mainly use chemical methods that include colloidal synthesis, sol-gel, and vapor phase deposition. The advantages of the bottom-up methods are the precise control over the size distribution and morphology of the resulting QDs, whereas the disadvantages are the extensive use of organic solvents, which produces a large quantity of harmful chemical waste and gas.

In the bottom-up approaches, there are a number of methods to synthesize QDs, including colloidal synthesis, sol-gel, and vapor phase deposition. Colloidal synthesis involves synthesizing QDs in an organic or inorganic solvent during a chemical reaction. Both the shape and size of the resulting QDs are affected by several factors, including temperature, reaction time, pH, addition rate of precursors, as well as concentrations of chemical reactants, and thus, can be controlled in order to attain the desired shape and size of QDs (Asatryan et al., 2016). This method is also known as the one-pot method, and it involves injecting semiconductor precursors rapidly into vigorously stirred hot organic solvents that contain molecules, which can coordinate with the surface of precipitated QDs.

There are three components in this synthesis, including surfactant molecules (ligands), dispersing phase, and dispersed phase (precursor). A precursor is typically converted into monomers as the reaction or colloidal state progresses, and when supersaturation is reached, nucleation begins. The nucleation and growth of QDs are governed by ligands that bind dynamically to the QD surfaces. The formation of undesirable sizes of crystals could also occur due to the dissolution of small-sized crystals as well as due to Ostwald ripening, which is the redeposition of small particles on larger particles. In this process, it can be observed that the smaller particles are dispersed in the solvent, while the larger particles are progressively growing. Furthermore, in order to ensure that the desired particle size is achieved, it is crucial to separate nucleation processes from the growing nanocrystals, and this can be done by controlling the temperature (Karakoti et al., 2015). In the initial stage of the reaction, heating is required to allow the precursor reagents to be converted into monomers and dispersed. It is then at a critical temperature when a second precursor is added at the specified speed rate to start nucleation. The reaction is then cooled promptly in order to halt the reaction as the nuclei develop into nanocrystals. The selection of ligands, reaction temperature, and reaction time allows precise control over the size, shape, and stoichiometry of QDs.

Sol-gel synthesis involves two phases, which are solution and gelation. "Sol" refers to the colloidal suspension of solid particles, and "gel" refers to the interconnected network of solid-phase particles. These solid-phase particles form a continuous structure of substances, which is typically a liquid phase. These phases are maintained through the chemical reactions that arise during gel evolutions and can be modified in various ways, including changing degree solvation, initial precursors, catalysts used, time for gelation, conditions for gelation, and physical processing of the gel itself. Sol-gel processes enable the formation of solid materials through the gelation of solutions (Chiriac et al., 2010).

The vapor phase deposition to synthesize QDs includes two main techniques, molecular beam epitaxy (MBE) and metalorganic chemical vapor deposition (MOCVD). MBE and MOCVD are favorable for growing QDs due to their fast growth rates. However, the growth temperatures of QDs are relatively high, above 1,000°C. Thus, low growth temperatures are challenging (Reilly et al., 2021).

8.3 THE SURFACE PROPERTIES OF QDS

As mentioned earlier, the formation of QDs can be described by the Stranski-Krastanov growth model. The quality of QDs is defined based on their size and shape uniformity, crystallinity, and completeness of surface passivation. The discrete electronic states of QDs are advantageous, as such highly monodispersed QDs can emit light and an intense narrow band or high color purity. In fluorescence imaging, however, the excitation light is irradiated into a large area of the target organ, and hence, tunable and unique light emission of QDs can distinguish them from the cellular environment. This allows single QD tracking in a cellular environment in fluorescence imaging (Zahid et al., 2018) and has also been implemented in fluorescence-guided surgery (Tian et al., 2020).

It is important to note that surface chemistry is responsible for modulating the optical properties of QDs. In particular, the change on the surface can alter the fluorescence intensity, excited state dynamics, and lifetime of the excited QDs. In addition, it is well known that new functionalities of QDs can be introduced through a subtle change on the surface, through which the interactions between QDs and surrounding molecules or substances can be controlled. This is the bottom line of the capability of QDs to be used as probes in detecting and labeling their surroundings. Therefore, in practical use, QDs have been utilized as probes for visualization, diagnostic, labeling, and therapeutic purposes, which are based on the interaction, bioimaging, and biosensing ability of QDs (Li & Zhu, 2013; Wagner et al., 2019). For these purposes, the surface of QDs should be treated and stabilized well by introducing proper capping agents, and are incorporated into living cells (Javed et al., 2020). At the same time, the biological, cytotoxicity, and optical properties of QDs are evaluated. In this sense, high-resolution QD images with resolution far beyond the diffraction limit of light could provide information at the molecular level.

As labeling agents, QDs capped with various capping agents having long-chain alkyl chains have been extensively utilized in fluorescent probes, photoelectrochemical assays, electrochemiluminescence, and optical transduction (Wagner et al., 2019). It is well known that capping agents having long-chain alkyl chains might prevent electron transfer so that they have a high fluorescence quantum yield. Compared with organic fluorophores, QDs show some advantages, including their photostability, narrow-band emission band, and quantum confinement effect. However, QDs contain heavy metals that are toxic and not biocompatible, so the biocompatibility of QDs relies on the capping agents. In this sense, the long-chain alkyl chains are hydrophobic, inert, and non-polar.

As mentioned earlier, the size that can be related to the surface-to-volume ratio of QDs is sensitive to their surrounding molecules or substances. Therefore, changing the QD size offers the manipulation of energy levels, drives the assembly of QDs, and is one of the routes to manipulate the interactions between QDs and their surrounding molecules or substances. In this sense, QDs can be bound to antibodies, proteins, or other biologic substances (Medintz et al., 2005). An extensive way to modify the QD surface is by designing proper capping agents so that they are conductive, facilitating charge transport and enhancing interdot coupling (Boles et al., 2016; Yazdani et al., 2020). The charge transport facilitates the charge carriers to cross the interparticle barrier, which could modify the interplay between electronic coupling, interfacial properties, and carrier confinement of QDs (Yazdani et al., 2020). The interdot coupling enhances the charge carrier mobility and shrinks the electronic gap. In this sense, the combination of their good mobility of photogenerated charge carriers and tunable absorption makes QDs potential optical sensors. Therefore, indium-based QDs have been demonstrated to be a good transport matrix and to induce an electric field, so they are a potential for sensors with high light sensitivity and low dark current (Lim et al., 2007; Martyniuk & Rogalski, 2008; Ren et al., 2019; Chen et al., 2020). The heterojunctions of QDs also provided charge circulation (Saran & Curry, 2016), enhancing infrared light sensitivity (Saran & Curry, 2016). Thus, not only are well-designed QDs with proper capping agents to passivate their surface fascinating advanced materials for bioprobes and bioassays, but at the same time, they can act as biosensors in analytic biochemistry assays.

8.4 QDS IN DRUG DELIVERY SYSTEMS

Biopolymers are capable of releasing the drug as they degrade, and they are then absorbed by the body. A drug therapy that targets a specific area of the body is ideal for achieving a certain amount of success. It should also be able to minimize the side effects of the medication. Various drug delivery systems, such as the controlled release of drugs, are being explored as drug carriers. One of the most important factors that can be considered when it comes to maintaining a constant drug concentration in the body is the rate at which the drug is released. This can be done through a single bolus injection. If the drug concentration is below the therapeutic range, then the pharmacological response can be limited. However, if the peak is greater than this range, then the side effects might occur. Various delivery systems have been studied to maintain a steady therapeutic level and minimize the side effects of the medication. These include the use of controlled-release techniques and sustained-release systems.

Traditional cancer treatment methods face many challenges, such as the time needed to develop and deliver multiple drugs and the difficulty in treating tumor sites. Single-drug therapy can lead to cancer reoccurrence and drug resistance. However, combining multiple therapies can improve the effectiveness of a treatment, but it can also cause side effects. Various drug carriers, including carbon nanomaterials, have been investigated (Banerjee et al., 2016). The earliest known drug carrier is the liposomes, which are nanoparticles based on solid lipids and dendrimer nanometer carriers (Liu, Chen, & Zhang, 2022). A new type of drug carrier that is capable of delivering drugs through the interaction of hydrophobic and hydrophilic groups is known as polymer micelles. These are formed by combining these two groups in water (Saxena et al., 2013).

Scientists are working on developing new drug carriers that can improve the effectiveness and safety of cancer treatment. They are currently developing a new type of chemical-siRNA combination therapy that can be used to treat various types of cancer. This method utilizes the targeted gene silencing of nanoparticles to compensate for the incomplete effects of chemotherapy. In addition, the development of new drug carriers using nanoparticles has been accelerated due to their potential to improve the effectiveness of cancer treatment. A study conducted by Banerjee et al. revealed that rod-like nanoparticles have better trans-intestinal cell transport and cell absorption than spherical nanoparticles (Banerjee et al., 2016). This finding paved the way for the rational design of nanodrugs for oral administration. Red blood cells are known to be one of the most natural drug carriers due to their long life and ability to carry breathing gas. The rapid emergence and evolution of nanotechnology has contributed to the development of new drug delivery systems. One of these is the nanoscale drug carriers, which are a type of nanometer-sized particles that can be used to deliver drugs. These systems can alter the speed of drug release and improve biodistribution in vivo. Nanoparticles are solid particles that range in size from a few nanometers to a few tens of nanometers, which are used as core components in functionalization systems. They can be used as carriers for drugs or as part of medical treatment. Currently, the size of nanoparticles used in medical applications is less than 100 nm.

Different preparation methods have been used for creating nanospheres and nanocapsules. A new type of solid lipid nanoparticle delivery system is continuously being developed. This type of drug carrier uses solid lipid as its carrier. It then wraps drugs in lipid nuclei to create solid particles. The chemical constituents of nanomaterials used in this process include gelatin, chitosan, carbon-based carriers, and branched polymers. Nanometer carriers for medical applications should not cause any negative effects or immune responses. The physiological environment can be changed to allow the drug-controlled release of nanoparticles. For instance, if the pH level of the environment changes, the drug-nanometer carrier can reach diseased tissues. To avoid injury to the body, the smaller nanoparticles can be easily removed by the kidney and tissue exosmosis. However, the larger ones can be cleared by the reticuloendothelial system.

QDs were the first nanotechnologies that were incorporated with biological applications, including their uses as photonic devices (Sugawara, Yamamoto, & Ebe, 2007), sensor materials (Zhang et al., 2005; Blasiak, van Veggel, & Tomanek, 2013), and fluorescent labels (Wagner et al., 2019) to replace toxic synthetic dyes in biomedical applications, including cancer therapy (Matea et al., 2017). QDs have also been utilized for detecting and treating diseases (Boriachek et al., 2017), bioimaging with high contrasts (Farias et al., 2018; Liu et al., 2016; Bilan et al., 2015; Ag et al., 2014; Kairdolf et al., 2013), drug delivery (Hafezi et al., 2021; Bindal, 2021; Rodriguez-Fragoso et al., 2014), as well as engineering tissues and biomarkers (Matea et al., 2017; Corredor et al., 2009; Santos et al., 2010). The main reason is that QDs are able to emit different wavelengths from visible to near-infrared spectral regions and, thus, enable nanocrystal QDs to be viewed as artificial atoms that can be modified to accommodate a specific technological application or a certain experiment (Klimov, 2015). As for representative examples, CdSe/ZnS QDs have been used to detect pathogens, such as *Escherichia coli* and *Hepatitis B* (Liu et al., 2007).

CdSe/ZnS QDs have also been used for tailoring the fluorescence of dental resin composite (Pilla et al., 2007). With their excellent optical properties, good photostability, and high chemical stability (Cheng et al., 2018), QDs have also been explored as fluorescent labels in order to replace toxic synthetic dyes in biomedical applications, such as in cancer therapies (Matea et al., 2017; Yuan, Hein, & Misra, 2010), diagnostics and therapeutic agents (Purushothaman & Song, 2021; Blasiak, van Veggel, & Tomanek, 2013; McMillan, Batrakova, & Gendelman, 2011), and disease detections (Fang et al., 2012; Goreham et al., 2019), and as long-circulating vascular markers (Walling et al., 2009; Smith et al., 2007) in reticuloendothelial system mapping (Hanaki et al., 2003), lymph node mapping (Kim et al., 2004), and anti-cancer applications (Ballou et al., 2007).

As mentioned in the previous section, the interest in QD application as drug nanocarriers is due mainly to its photostable fluorescent properties (Liu, Brandon, & Cate, 2007; Pilla et al., 2007), facilitating traceable drug delivery. Therefore, most of the studies in this field have been focused on the labeling of conventional drug carriers with QDs or on the functionalization of QDs with π-conjugated organic compounds (Shahri et al., 2022). Considering that poly(lactic-co-glycolic acid) and polyethyleneimine are among the polymers used as drug carriers, biopolymers have been mostly attracted as capping agents; for instance, to replace small thiol compounds to passivate QDs and to overcome the limitation associated with for real-time and long-term imaging and tracking of drug transport (Lim et al., 2021). QDs have also been potentially labeled with both inorganic and organic drug carriers, or even with bacteria and viruses (Chen et al., 2005; Tan, Jiang, & Zhang, 2007; Jia et al., 2007; Akin et al., 2006). Their excellent fluorescent properties make QDs useful in cellular membrane delivery, as they are able to penetrate into the nucleus (Al-Nahain et al., 2013).

It is also interesting to note that QDs are introduced and tagged into the cytoplasm of cells or the nucleus, which is the most sensitive to damage. Therefore, in addition to their excellent fluorescent properties, cytotoxicity is another characteristic of QDs. The bulk semiconductors and their QDs are found to be acutely toxic under certain conditions (Derfus, Chan, & Bhatia, 2004), while their toxicity is reduced when they are encapsulated into the core-shell formation using ZnS as a shell (Hardman, 2006). With the non-toxic ZnS, CdSe/ZnS QDs have been demonstrated to induce no cell damage, revealing that the capping technique implementing proper capping agents and surface coatings is an excellent approach to suppress the cytotoxicity of QDs.

It is interesting to recall that capping agents can be designed to modify QD properties. This allows QDs to be useful as drug labels and vehicles in diagnostic, therapeutic, and drug delivery systems. In particular, the design of capping agents is important to control not only the size, morphology, and surface chemistry but also the functionality and biological reactivity and functionality of QDs. Therefore, QDs have been explored for biological and biomedicinal applications (Anderson, Gwenin, & Gwenin, 2019; Römer et al., 2019). The most crucial consideration is designing capping agents so that the agglomeration, growth, and physicochemical properties of QDs are precisely controlled (Niu & Li, 2014). Moreover, the resulting QDs are useful as drug vehicles in diagnostic, therapeutic, and drug delivery systems because of their non-toxicity and various functional groups, which could readily interact with those of drugs. It is well known that capping agents should be amphiphilic, consisting of polar and non-polar moieties, so that while passivating the metal atoms of the QD surface, the capping agents could enhance the functionality and the compatibility with surrounding environments (Gulati, Sachdeva, & Bhasin, 2018). There are some possible potential capping agents, such as small ligands, surfactants, dendrimers, synthetic inert polymers, cyclodextrins, biopolymers, and aqueous plant extracts. These capping agents have been successfully utilized as capping agents of various QDs. Exploring these potential capping agents opens up more possibilities for designing QDs with good biological functionalities and ameliorated colloidal stability for biological and biomedicinal applications. Several of the potential capping agents are described subsequently.

8.4.1 POLYETHYLENE GLYCOL

Polyethylene glycol (PEG) is a biocompatible, non-immunogenic, and non-toxic synthetic polymer. Its hydroxyl groups make this polymer hydrophilic and soluble in water and organic solvents. Water affinity and biodegradability of PEG are granted by its hydrophilic characteristics (Shameli et al., 2012). Therefore, PEG has been utilized as a capping agent to stabilize various biocompatible Au,

Ag, and Zn nanoparticles, which are useful as drug carriers (Zamora-Justo et al., 2019; Pinzaru et al., 2018; Singletary et al., 2017).

8.4.2 POLYVINYL ALCOHOL

Polyvinyl alcohol (PVA) has a good capping agent performance and great hydrophilicity. PVA is essentially biocompatible and biodegradable, and has been utilized to stabilize Ag nanoparticles (Kyrychenko, Pasko, & Kalugin, 2017). PVA has been utilized in various biomedical applications, and PVA hydrogels show their potential for artificial implants and tissue engineering (Gaaz et al., 2015). PVA has also been used as a capping agent for iron oxide nanoparticles (Rahayu et al., 2017).

8.4.3 POLYVINYLPYRROLIDONE

Polyvinylpyrrolidone (PVP) is composed of N-vinylpyrrolidone monomer and is biocompatible, chemically stable, non-toxic, and soluble in aqueous solution and organic solvents. PVP forms a compelling coating agent, so it is a potential polymer for many considerable non-medical and medical purposes. In this sense, PVP has been a common ingredient in pellets, capsules, tablets, granules, syrups, and injectable solutions, and has been used as coating of contact lenses (Teodorescu & Bercea, 2015; Okoroh et al., 2019). PVP has also been applied as a capping agent to stabilize various biocompatible Fe, Au, Ag, and Zn nanoparticles (Ahlberg et al., 2014).

8.4.4 BOVINE SERUM ALBUMIN

Bovine serum albumin (BSA) is a ubiquitous and well-characterized protein in mammals. BSA has charged carboxyl, sulfhydryl, and amino groups, and thus, has the ability to bind various therapeutic systems, including poly-conjugated dyes and drugs. BSA has also been applied as a capping agent, and BSA-capped QDs are highly bioavailable. BSA can be used as a source of nutrients and amino acids, so that would be taken up by cells, including tumor cells. This offers direct applications of BSA-capped QDs to treat tumor cells. BSA has also been applied as capping agents of Pd-, Au-, Pt-, and Ag-based nanoparticles (Au et al., 2010; Bolaños, Kogan, & Araya, 2019). In particular, BSA can act as a structure-directing agent, controlling the nucleation, assembly, and growth of nanoparticles. BSA-capped Au nanoparticles are highly stable and can be easily conjugated (Bolaños, Kogan, & Araya, 2019), and BSA-capped CdS QDs have been developed as a sensor for the detection of protamine and heparin (Li & Yang, 2015).

8.4.5 ETHYLENE DIAMINE TETRA ACETIC ACID

Ethylene diamine tetra acetic acid (EDTA) is soluble in water and is commonly utilized as a chelating agent. EDTA is good for interacting with divalent metal ions and has been an important capping agent to stabilize Au, Zn, Cd, Cr, and Cu nanoparticles (Harish & Reddy, 2015; Reddy et al., 2011). EDTA has attracted great interest as a stabilizer in the fabrication of nanoparticles and QDs.

8.4.6 CHITIN AND CHITOSAN

Chitin and chitosan are biopolymers consisting of N-acetyl-D-glucosamine and D-glucosamine derived from the exoskeleton of crustaceans, arachnids, insects, and fungi. Chitin and chitosan are the second most abundant biopolymers in nature. The acetyl, amino, and hydroxyl groups of chitin and chitosan are important functional groups, as they could form intermolecular hydrogen bonding interactions with a wide range of organic compounds, including drugs. The conformation of chitin and chitosan is flexible. Chitin and chitosan are biocompatible, biodegradable, antimicrobial, immunoenhancing, anti-carcinogenic, and non-toxic. Chitin and chitosan are soluble biopolymers in slightly acidic aqueous solutions (Elgadir et al., 2015). Chitosan, especially, has attracted great attention owing to its exceptional biological properties, so this biopolymer has been useful as advanced material for biomedical applications. In nanoscience and nanotechnology, chitin and chitosan have been extensively utilized as capping agents of metal nanoparticles by the wet-chemical synthesis method. Though the bonding between chitin and chitosan and the metallic atoms is still unclear, it is believed that the acetyl, amino, and hydroxyl groups of chitin and chitosan coordinate the multivalent metallic atoms on the surface of nanoparticles (Deng et al., 2017). Therefore, chitosan has been applied to control the size, morphology, and optical properties, as well as to stabilize Au, Ag, and Cu nanoparticles (Franconetti et al., 2019; Cinteza et al., 2018; Jayaramudu et al., 2019). Chitin- and chitosan-capped CdS QDs have been synthesized for drug carriers (Lim et al., 2021; Abdelhamid et al., 2019).

8.4.7 PLANT EXTRACTS

Plant extracts have also been of interest in the fabrication of nanoparticles, as some compounds of the extracts have potential as reducing agents. The compounds in the extracts can influence the physical and chemical properties of nanoparticles (Aisida, Ugwu, Akpa, Nwanya, Ejikeme et al., 2019; Aisida, Ugwu, Akpa, Nwanya, Nwankwo et al., 2019; Aisida, Madubuonu, Alnasir et al., 2020). A large number of works have been devoted to finding functional nanoparticles for biomedical applications (Aisida Ugwu, Akpa, Nwanya, Ejikeme et al., 2019; Aisida Ugwu, Akpa, Nwanya, Nwankwo et al., 2019; Aisida, Ahmad, Zhao et al., 2020).

8.5 FUTURE OF QDS IN BIOMEDICAL APPLICATIONS

The use of QDs in nanoscience and nanotechnology should continue to expand. In particular, research efforts have successfully applied QDs in broad optical and optoelectronic, photoluminescence, and photovoltaic applications. QDs have been developed in optical and semiconducting fields using proper capping agents. As QDs are suitable to label live cells, they can be used to measure cell motility as well as the pharmacokinetics and pharmacodynamics of drugs. The cells are capable of allowing and overcoming the insertion of QDs. Systemic injection of QDs as nanoprobes can provide multicolor fluorescence imaging of cancer cells.

The ability of QDs to be used as traceable and efficient drug nanocarriers in drug delivery systems has prompted research to continue with the development of this approach. The distinct designable structure and surface properties of QDs enable their usage in targeted drug delivery. The possibility to design new types of specific structural features of capping agents to enhance the functionality and the compatibility of QDs with biological substances has brought about the benefits of this advanced material in biological and biomedical areas. In particular, QDs are useful for visualization, diagnostic, labeling, and therapeutic purposes. Possible utilization of QDs in new development areas, including biological and biomedical applications, depends on the design of proper capping agents (Klostranec & Chan, 2006). In this sense, capping agents to be explored in the future include various organic compounds, amino acids, and polymers. These types of capping agents will change the face of QD technologies for a better future and are expected to promote QDs in many biomedical disciplines, such as cancer diagnostics, stem cell therapeutics, and embryogenesis.

8.6 CONCLUSIONS

This chapter summarizes several important characteristics of quantum dot (QD) as one of the advanced and forefront materials, and several aspects of its synthesis and conjugation. Capping agents that prevent the uncontrolled growth and aggregation of small-sized clusters and the formation of semiconductor nanocrystals into larger-size particles have opened up an extensive way to modify the QD surface so that QDs could facilitate efficient charge transport and interdot coupling. Monodispersed QDs are crucial in determining the vital applications of QDs. Due to their distinctive photostability, distinctive design and engineering of capping agents, and rich surface chemistry, QDs can be used as traceable drug nanocarriers in drug delivery systems. Specific structural features of capping agents can also be used to alter the biological activities of QDs for visualization, diagnostic, labeling, and therapeutic purposes. Designing the specific structural features of capping agents can enhance the functionality and the compatibility of QDs with biological substances, and has brought about the benefits of this advanced material for visualization, diagnostic, labeling, and therapeutic purposes. Various capping agents, including organic compounds, amino acids, and polymers, could be explored in the future, and they are expected to change the face of QD technologies for a better future, promoting QDs for many biomedical disciplines, such as cancer diagnostics, stem cell therapeutics, and embryogenesis.

REFERENCES

Abdelhamid HN, El-Bery HM, Metwally AA, Elshazly M, Hathout RM. 2019. Synthesis of CdS-modified chitosan quantum dots for the drug delivery of sesamol. *Carbohydr. Polym.* 214, 90–99

Aisida SO, Ahmad I, Zhao T-K, Maaza M, Ezema FI. 2020. Calcination effect on the photoluminescence, optical, structural, and magnetic properties of polyvinyl alcohol doped ZnFe2O4 nanoparticles. *J. Macromol. Sci. Part B* 59, 295–308.

Aisida SO, Madubuonu N, Alnasir MH, Ahmad I, Botha SS, Maaza M, Ezema FI. 2020. Biogenic synthesis of iron oxide nanorods using *Moringa oleifera* leaf extract for antibacterial applications. *Appl. Nanosci.* 10, 305–315.

Aisida SO, Ugwu K, Akpa PA, Nwanya AC, Ejikeme PM, Botha SS, Ahmad I, Maaza M, Ezema FI. 2019. Biogenic synthesis and antibacterial activity of controlled silver nanoparticles using an extract of *Gongronema latifolium*. *Mater. Chem. Phys.* 237, 121859.

Aisida SO, Ugwu K, Akpa PA, Nwanya AC, Nwankwo U, Botha SS, Ejikeme PM, Ahmad I, Maaza M, Ezema FI. 2019. Biosynthesis of silver nanoparticles using bitter leave (*Veronica amygdalina*) for antibacterial activities. *Surf. Interfaces* 17, 100359.

Ag D, Bongartz R, Dogan LE, Seleci M, Walter JG, Demirkol DO, Stahl F, Ozcelik S, Timur S, Scheper T. 2014. Biofunctional quantum dots as fluorescence probe for cell-specific targeting. *Colloid. Surf. B: Biointerfaces* 114, 96–103.

Ahlberg S, Antonopulos A, Diendorf J, Dringen R, Epple M, Flöck R, Goedecke W, Graf C, Haberl N, Helmlinger J, Herzog F, Heuer F, Hirn S, Johannes C, Kittler S, Köller M, Korn K, Kreyling WG, Krombach F, Lademann J, Loza K, Luther EM, Malissek M, Meinke MC, Nordmeyer D, Pailliart A, Raabe J, Rancan F, Rothen-Rutishauser B, Rühl E, Schleh C, Seibel A, Sengstock C, Treuel L, Vogt A, Weber K, Zellner R. 2014. PVP-coated, negatively charged silver nanoparticles: A multi-center study of their physicochemical characteristics, cell culture and in vivo experiments. *Beilstein J. Nanotechnol.* 5, 1944–1965.

Akin D, Sturgis J, Ragheb K, Sherman D, Burkholder K, Robinson JP, Bhunia AK, Sulma Mohammed S, Bashir R. 2006. Bacteria-mediated delivery of nanoparticles and cargo into cells. *Nat. Nanotechnol.* 2, 441–449.

Alivisatos AP. 1996. Semiconductor clusters, nanocrystals and quantum dots. *Science* 271 (5251), 933–937.

Al-Nahain A, Lee JE, In I, Lee H, Lee KD, Jeong JH, Park SY. 2013. Target delivery and cell imaging using hyaluronic acid-functionalized graphene quantum dots. *Mol. Pharmaceutics* 10, 3736–3744.

Anderson SD, Gwenin VV, Gwenin CD. 2019. Magnetic functionalized nanoparticles for biomedical, drug delivery and imaging applications. *Nanoscale Res. Lett.* 14, 188.

Antolini F, Orazi L. 2019. Quantum dots synthesis through direct laser patterning: A review. *Front. Chem.* 7, 252.

Apter B, Lapshina N, Handelman A, Fainberg BD, Rosenman G. 2018. Peptide nanophotonics: From optical waveguiding to precise medicine and multifunctional biochips. *Small* 14 (34), e1801147.

Asatryan AL, Vartanian A, Kirakosyan AA, Vardanyan LA. 2016. Electric field and image charge effects on impurity-bound polarons in a CdS colloidal quantum dot embedded in organic matrices. *Phys. B Condens. Matter* 503, 70–74.

Au L, Lim B, Colletti P, Jun YS, Xia Y. 2010. Synthesis of gold microplates using bovine serum albumin as a reductant and a stabilizer. *Chem. Asian J.* 5 (1), 123–129.

Ballou B, Ernst L, Andreko S, Harper T, Fitzpatrick J, Waggoner A, Bruchez M. 2007. Sentinel lymph node imaging using quantum dots in mouse tumor models. *Bioconjug. Chem.* 18 (2), 389–396.

Banerjee A, Qi J, Gogoi R, Wong J, Mitragotri S. 2016. Role of nanoparticle size, shape and surface chemistry in oral drug delivery. *J. Control Release* 238, 176–185.

Baqalkot V, Zhang L, Levy-Nissenbaum E, Jon S, Kantoff PW, Langer R, Farokhzad OC. 2007. Quantum dot-aptamer conjugates for synchronous cancer imaging, therapy, and sensing of drug delivery based on bi-fluorescence resonance energy transfer. *Nano Lett.* 7 (10), 3065–3070.

Bilan R, Fleury F, Nabiev I, Sukhanova A. 2015. Quantum dot surface chemistry and functionalization for cell targeting and imaging. *Bioconjug. Chem.* 26 (4), 609–624.

Bindal C. 2021. A novel multifunctional NCQDs-based injectable self-crosslinking and in situ forming hydrogel as an innovative stimuli responsive smart drug delivery system for cancer therapy. *Mater. Sci. Eng. C* 121, 111829.

Biswas A, Bayer IS, Biris AS, Wang T, Dervishi E, Faupel F. 2012. Advances in top-down and bottom-up surface nanofabrication: Techniques, applications & future prospects. *Adv. Colloid Interface Sci.* 170 (1–2), 2–27.

Blasiak B, van Veggel FCJM, Tomanek B. 2013. Applications of nanoparticles for MRI cancer diagnosis and therapy. *J. Nanomater.* 2013, 148578.

Bolaños K, Kogan MJ, Araya E. 2019. Capping gold nanoparticles with albumin to improve their biomedical properties. *Int. J. Nanomed.* 14, 6387–6406.

Boles MA, Ling D, Hyeon T, Talapin DV. 2016. The surface science of nanocrystals. *Nat. Mater.* 15, 141–153.

Boriachek K, Islam MN, Gopalan V, Lam AK, Nguyen NT, Shiddiky MJA. 2017. Quantum dot-based sensitive detection of disease specific exosome in serum. *Analyst* 142 (12), 2211–2219.

Chen AA, Derfus AM, Khetani SR, Bhatia SN. 2005. Quantum dots to monitor RNAi delivery and improve gene silencing. *Nucleic Acids Res.* 33 (22), e190.

Chen B, Wan Y, Xie Z, Huang J, Zhang N, Shang C, Norman J, Li Q, Tong Y, Lau KM, Gossard AC, Bowers JE. 2020. Low dark current high gain InAs quantum dot avalanche photodiodes monolithically grown on Si. *ACS Photonics* 7, 528–533.

Cheng L, Xiang Q, Liao Y, Zhang H. 2018. CdS-based photocatalysts. *Energy Environ. Sci.* 11 (6), 1362–1391.

Chiriac A, Neamtu I, Nita L, Nistor M. 2010. Sol-gel method performed for biomedical products implementation. *Mini Rev. Med. Chem.* 77, 990–1013.

Cinteza LO, Scomoroscenco C, Voicu SN, Nistor CL, Nitu SG, Trica B, Jecu ML, Petcu C. 2018. Chitosan-stabilized ag nanoparticles with superior biocompatibility and their synergistic antibacterial effect in mixtures with essential oils. *Nanomaterials* 8 (10), 826.

Corredor E, Testillano P, Coronado M, González-Melendi P, Pacheco R, Marquina C, Ibarra M, Fuente J, Rubiales D, Pérez-de-Luque A, Risueño MC. 2009. Nanoparticle penetration and transport in living pumpkin plants: In situ subcellular identification. *BMC Plant Biol.* 9, 45.

de Arquer FPG, Talapin DV, Klimov VI, Arakawa Y, Bayer M, Sargent EH. 2021. Semiconductor quantum dots: Technological progress and future challenges. *Science* 373, eaaz8541.

Deng HH, Lin XL, Liu YH, Li KL, Zhuang QQ, Peng HP, Liu AL, Xia XH, Chen W. 2017. Chitosan stabilized platinum nanoparticles as effective oxidase mimics for colorimetric detection of acid phosphatase. *Nanoscale* 9, 10292–10300.

Derfus AM, Chan WCW, Bhatia SN. 2004. Probing the cytotoxicity of semiconductor quantum dots. *Nano Lett.* 4, 11–18.

Ekimov AI, Onushchenko AA. 1981. Quantum size effect in three-dimensional microscopic semiconductor crystals. *JETP Lett.* 34 (6), 345–349.

Ekimov EI, Efros AL, Onushchenko AA. 1985. Quantum size effect in semiconductor microcrystals. *Solid State Commun.* 56 (11), 921–924.

Elgadir A, Uddin MS, Ferdosh S, Adam S, Chowdhury AJK, Sarker MZI. 2015. Impact of chitosan composites and chitosan nanoparticle composites on various drug delivery systems: A review. *J. Food Drug Anal.* 23, 619–629.

Fang M, Peng C, Pang D, Li Y. 2012. Quantum dots for cancer research: Current status, remaining issues, and future perspectives characteristics of QDs for biomedical. *Cancer Biol. Med.* 9 (3), 151–163.

Farias PMA, Galembeck A, Mendonca WS, Stingl A. 2018. The colloidal quantum dots suitability for long term cell imaging. *J. Adv. Chem.* 15 (2), 6275–6281.

Franconetti A, Carnerero JM, Prado-Gotor R, Cabrera-Escribano F, Jaime C. 2019. Chitosan as a capping agent: Insights on the stabilization of gold nanoparticles. *Carbohydr. Polym.* 207, 806–814.

Gaaz TS, Sulong AB, Akhtar MN, Kadhum AAH, Mohamad AB, Al-Amiery AA, 2015. Properties and applications of polyvinyl alcohol, halloysite nanotubes and their nanocomposites. *Molecules* 20 (12), 22833–22847.

Goreham RV, Ayed Z, Ayupova D, Dobhal G. 2019. Extracellular vesicles: Nature's own nanoparticles. *Compr. Nanosci. Nanotechnol.* 1–5, 27–48.

Gulati S, Sachdeva M, Bhasin KK. 2018. Capping agents in nanoparticle synthesis: Surfactant and solvent system. *AIP Conf. Proc.* 1953, 030214.

Hafezi M, Rostami M, Hosseini A, Rahimi-Nasrabadi M, Fasihi-Ramandi M. 2021. Cur-loaded $ZnFe_2O_4$@mZnO @N-GQDs biocompatible nano-carriers for smart and controlled targeted drug delivery with pH-triggered and ultrasound irradiation. *J. Mol. Liq.* 322, 114875.

Hanaki KI, Momo A, Oku T, Komoto A, Maenosono S, Yamaguchi Y, Yamamoto K. 2003. Semiconductor quantum dot/albumin complex is a long-life and highly photostable endosome marker. *Biochem. Biophys. Res. Commun.* 302 (3), 496–501.

Hardman R. 2006. A toxicologic review of quantum dots: toxicity depends on physicochemical and environmental factors. *Environ. Health Perspect.* 114, 165–172.

Harish GS, Reddy PS. 2015. Synthesis and characterization of water soluble ZnS: Ce, Cu co-doped nanoparticles: Effect of EDTA concentration. *Int. J. Sci. Res.* 4, 221–225.

Horoz S, Lu L, Dai Q, Chen J, Yakami B, Pikal JM, Wang W, Tang J. 2012. CdSe quantum dots synthesized by laser ablation in water and their photovoltaic applications. *Appl. Phys. Lett.* 101, 223902.

Hu R, Law WC, Lin G, Ye L, Liu J, Liu J, Reynolds JL, Yong KT. 2012. PEGylated phospholipid micelle-encapsulated near-infrared PbS quantum dots for in vitro and in vivo bioimaging. *Theranostics* 2 (7), 723–733.

Javed R, Zia M, Naz S, Aisida SO, Ain N, Ao Q. 2020. Role of capping agents in the application of nanoparticles in biomedicine and environmental remediation: Recent trends and future prospects. *J. Nanobiotechnol.* 18, 172.

Jayaramudu T, Varaprasad K, Pyarasani RD, Reddy KK, Kumar KD, Akbari-Fakhrabadi A, Mangalaraja RV, Amalraj J. 2019. Chitosan capped copper oxide/copper nanoparticles encapsulated microbial resistant nanocomposite films. *Int. J. Biol. Macromol.* 128, 499–508.

Jia N, Lian Q, Shen H, Wang C, Li X, Yang Z. 2007. Intracellular delivery of quantum dots tagged antisense oligodeoxynucleotides by functionalized multiwalled carbon nanotubes. *Nano Lett.* 7 (10), 2976–2980.

Kairdolf BA, Smith AM, Stokes TH, Wang MD, Young AN, Nie S. 2013. Semiconductor quantum dots for bioimaging and biodiagnostic applications. *Annu. Rev. Anal. Chem.* 6 (1), 143–162.

Kandasamy K, Surendhiran S, Syed Khadar YA, Rajasingh P. 2021. Ultrasound-assisted microwave synthesis of CdS/MWCNTs QDs: A material for photocatalytic and corrosion inhibition activity. *Mater Today Proc.* 47 (3), 757–762.

Karakoti AS, Shukla R, Shanker R, Singh S. 2015. Surface functionalization of quantum dots for biological applications. *Adv. Colloid Interface Sci.* 215, 28–45.

Kim S, Lim Y, Soltesz E, Grand A, Lee J, Nakayama A, Parker J, Mihaljevic T, Laurence R, Dor D, Cohn L, Bawendi M, Frangioni J. 2004. Near-infrared fluorescent type II quantum dots for sentinel lymph node mapping. *Nat. Biotechnol.* 22 (1), 93–97.

Klimov VI. 2015. Nanocrystal quantum dots. *Physiol. Res.* 64 (6), 897–905.

Klostranec, Chan WCW. 2006. Quantum dots in biological and biomedical research: Recent progress and present challenges. *Adv. Mater.* 18 (15), 1953–1964.

Kyrychenko A, Pasko DA, Kalugin ON. 2017. Poly(vinyl alcohol) as a water protecting agent for silver nanoparticles: The role of polymer size and structure. *Phys. Chem. Chem. Phys.* 19, 8742–8756.

Li H, Yang X. 2015. Bovine serum albumin-capped CdS quantum dots as an inner-filter effect sensor for rapid detection and quantification of protamine and heparin. *Anal. Methods* 7, 8445–8452.

Li J, Zhu JJ. 2013. Quantum dots for fluorescent biosensing and bio-imaging applications. *Analyst* 138, 2506–2515.

Li Y, Li Z, Wang X, Liu F, Cheng Y, Zhang B, Shi D. 2012. In vivo cancer targeting and imaging-guided surgery with near infrared-emitting quantum dot bioconjugates. *Theranostics* 2, 769–776.

Lim H, Tsao S, Zhang W, Razeghi M. 2007. High-performance InAs quantum-dot infrared photodetectors grown on InP substrate operating at room temperature. *Appl. Phys. Lett.* 90, 131112.

Lim MJ, Shahri NNM, Taha H, Mahadi AH, Kusrini E, Lim JW, Usman A. 2021. Biocompatible chitin-encapsulated CdS quantum dots: Fabrication and antibacterial screening. *Carbohydr. Polym.* 260, 117806.

Liu M, Xu Y, Niu F, Gooding JJ, Liu J. 2016. Carbon quantum dots directly generated from electrochemical oxidation of graphite electrodes in alkaline alcohols and the applications for specific ferric ion detection and cell imaging. *Analyst* 141, 2657–2664.

Liu P, Chen G, Zhang J. 2022. A review of liposomes as a drug delivery system: Current status of approved products, regulatory environments, and future perspectives. *Molecules* 27 (4): 1372.

Liu Y, Brandon R, Cate M, Peng X, Stony R, Michael Johnson M. 2007. Detection of pathogens using luminescent CdSe/ZnS dendron nanocrystals and a porous membrane immunofilter. *Anal. Chem.* 79 (22), 8796–8802.

Manabe N, Hoshino A, Liang YQ, Goto T, Kato N, Yamamoto K. 2006. Quantum dot as a drug tracer in vivo. *IEEE Trans. Nanobiosci.* 5 (4), 263–267.

Martyniuk P, Rogalski A. 2008. Quantum-dot infrared photodetectors: Status and outlook. *Prog. Quantum Electron.* 32, 89–120.

Matea CT, Mocan T, Pop T, Puia C, Iancu C. 2017. Quantum dots in imaging, drug delivery and sensor applications. *Int. J. Nanomed.* 12, 5421–5431.

McMillan J, Batrakova E, Gendelman HE. 2011. Cell delivery of therapeutic nanoparticles Prog. *Mol. Biol. Transl. Sci.* 104, 563–601.

Medintz IL, Uyeda HT, Goldman ER, Mattoussi H, 2005. Quantum dot bioconjugates for imaging, labelling and sensing. *Nat. Mater.* 4, 435–446.

Niu Z, Li Y. 2014. Removal and utilization of capping agents in nanocatalysis. *Chem. Mater.* 26 (1), 72–83.

Okoroh DO, Ozuomba JO, Aisida SO, Asogwa PU. 2019. Properties of zinc ferrite nanoparticles due to PVP mediation and annealing at 500°C. *Adv. Nanoparticles* 8, 36–45.

Pinaud F, Michalet X, Bentolila LA, Tsay JM, Doose S, Li JJ, Iyer G, Weiss S. 2006. Advances in fluorescence imaging with quantum dot bio-probes. *Biomaterials* 27, 1679–1687.

Pilla V, Alves LP, Munin E, Pacheco MTT. 2007. Radiative quantum efficiency of CdSe/ZnS quantum dots suspended in different solvents. *Opt. Commun.* 280 (1), 225–229.

Pinzaru I, Coricovac D, Dehelean C, Moacă EA, Mioc M, Baderca F, Sizemore I, Brittle S, Marti D, Calina CD, Tsatsakis AM, Şoica C. 2018. Stable PEG-coated silver nanoparticles – a comprehensive toxicological profile. *Food Chem. Toxicol.* 111, 546–556.

Pisanic TR II, Zhang Y, Wang TH. 2014. Quantum dots in diagnostics and detection: Principles and paradigms. *Analyst* 139, 2968–2981.

Protesescu L, Yakunin S, Nazarenko O, Dirin DN, Maksym V. Kovalenko MV. 2018. Low-cost synthesis of highly luminescent colloidal lead halide perovskite nanocrystals by wet ball milling. *ACS Appl. Nano Mater.* 1 (3), 1300–1308.

Purushothaman B, Song JM. 2021. Ag$_2$S quantum dot theragnostics. *Biomater. Sci.* 9, 51–69.

Rahayu LBH, Wulandari IO, Santjojo DH, Sabarudin A. 2017. Synthesis and characterization of Fe$_3$O$_4$ nanoparticles using polyvinyl alcohol (PVA) as capping agent and glutaraldehyde (GA) as crosslinker. *IOP Conf. Ser. Mater. Sci. Eng.* 299, 012062.

Reddy DA, Murali G, Vijayalakshmi RP, Reddy BK. 2011. Room-temperature ferromagnetism in EDTA capped Cr-doped ZnS nanoparticles. *Appl. Phys. A Mater. Sci. Process* 105, 119–124.

Reilly CE, Keller S, Nakamura S, DenBaars SP. 2021. Metalorganic chemical vapor deposition of InN quantum dots and nanostructures. *Light Sci. Appl.* 10, 150.

Ren A, Yuan L, Xu H, Wu J, Wang Z. 2019. Recent progress of III-V quantum dots infrared photodetectors on silicon. *J. Mater. Chem. C Mater. Opt. Electron. Devices* 7, 14441–14453.

Rodriguez-Fragoso L, Gutiérrez-Sancha I, Rodríguez-Fragoso P, Rodríguez-López A, Reyes-Esparza J. 2014. Pharmacokinetic properties and safety of cadmium-containing quantum dots as drug delivery systems; In: A. D. Sezer (Ed.), *Application of Nanotechnology in Drug Delivery*, IntechOpen, Rijeka, Croatia, Chapter 14. https://www.intechopen.com/chapters/46953

Römer I, Briffa SM, Dasilva YAR, Hapiuk D, Trouillet V, Palmer RE, Valsami-Jones E. 2019. Impact of particle size, oxidation state and capping agent of different cerium dioxide nanoparticles on the phosphate-induced transformations at different pH and concentration. *PLoS ONE* 14, 0217483.

Santos AR, Miguel AS, Tomaz L, Malhó R, Maycock C, Vaz Patto MC, Fevereiro P, Oliva A. 2010. The impact of CdSe/ZnS quantum dots in cells of medicago sativa in suspension culture. *J. Nanobiotechnol.* 8 (1), 24.

Saran R, Curry RJ. 2016. Lead sulphide nanocrystal photodetector technologies. *Nat. Photonics* 10, 81–92.

Saxena V, Diaz A, Clearfield A, Batteas JD, Hussain MD. 2013. Zirconium phosphate nanoplatelets: A biocompatible nanomaterial for drug delivery to cancer. *Nanoscale* 5 (6), 2328–2336.

Shahri NNM, Taha H, Hamid MHSA, Kusrini E, Lim JW, Hobley J, Usman A. 2022. Antimicrobial activity of silver sulfide quantum dots functionalized with highly conjugated Schiff bases in a one-step synthesis. *RSC Adv.* 12, 3136–3146.

Shameli K, Ahmad MB, Zamanian A, Sangpour P, Shabanzadeh P, Abdollahi Y, Zargar M. 2012. Green biosynthesis of silver nanoparticles using Curcuma longa tuber powder. *Int. J. Nanomed.* 7, 5603–5610.

Singletary M, Hagerty S, Muramoto S, Daniels Y, MacCrehan WA, Stan G, Lau JW, Pustovyy O, Globa L, Morrison EE, Sorokulova I, Vodyanoy V. 2017. PEGylation of zinc nanoparticles amplifies their ability to enhance olfactory responses to odorant. *PLoS ONE* 12, 0189273.

Smith JD, Fisher GW, Waggoner AS, Campbell PG. 2007. The use of quantum dots for analysis of chick CAM vasculature. *Microvasc. Res.* 73 (2), 75–83.

Suarez JA, Plata JJ, Marquez AM, Sanz JF. 2017. Effects of the capping ligands, linkers and oxide surface on the electron injection mechanism of copper sulfide quantum dot-sensitized solar cells. *Phys. Chem. Chem. Phys.* 19, 14580–14587.

Sugawara M, Yamamoto T, Ebe H. 2007. Quantum-dot-based photonic devices. *FUJITSU Sci. Tech. J.* 43, 495–501.

Takezawa K, Lu J, Numako C, Takami S. 2021. One-step solvothermal synthesis and growth mechanism of well-crystallized β-Ga2O3 nanoparticles in isopropanol. *CrystEngComm* 23, 6567.

Tan WB, Jiang S, Zhang Y. 2007. Quantum-dot based nanoparticles for targeted silencing of HER2/NEU gene via RNA interference. *Biomaterials* 28 (8), 1565–1571

Teodorescu M, Bercea M. 2015. Poly(vinylpyrrolidone) – a versatile polymer for biomedical and beyond medical applications. *Polym. Plast. Technol. Eng.* 54, 923–943.

Tian R, Ma H, Zhu S, Lau J, Ma R, Liu Y, Lin L, Chandra S, Wang S, Zhu X, Deng H, Niu G, Zhang M, Antaris AL, Hettie KS, Yang B, Liang Y, Chen X. 2020. Multiplexed NIR-II probes for lymph node-invaded cancer detection and imaging-guided surgery. *Adv. Mater.* 32, 1907365.

Valizadeh A, Mikaeili H, Samiei M. 2012. Quantum dots: Synthesis, bioapplications, and toxicity. *Nanoscale Res. Lett.* 7 (1), 480.

Wagner AM, Knipe JM, Orive G, Peppas NA. 2019. Quantum dots in biomedical applications. *Acta Biomater.* 94, 44–63.

Walling MA, Novak JA, Shepard JRE. 2009. Quantum dots for live cell and in vivo imaging. *Int. J. Mol. Sci.* 10 (2), 441–491.

Yang G, Park SJ. 2019. Conventional and microwave hydrothermal synthesis and application of functional materials: A review. *Materials* 12 (7), 1177.

Yazdani N, Andermatt S, Yarema M, Farto V, Bani-Hashemian MH, Volk S, Lin WMM, Yarema O, Mathieu Luisier M, Wood V. 2020. Charge transport in semiconductors assembled from nanocrystals. *Nat. Commun.* 11, 2852.

Yuan Q, Hein S, Misra RDK. 2010. New generation of chitosan-encapsulated ZnO quantum dots loaded with drug: Synthesis, characterization and in vitro drug delivery response. *Acta Biomater.* 6 (7), 2732–2739.

Zahid MU, Ma L, Lim SJ, Smith AM. 2018. Single quantum dot tracking reveals the impact of nanoparticle surface on intracellular state. *Nat. Commun.* 9, 1830.

Zamora-Justo JA, Abrica-González P, Vázquez-Martínez RG, Muñoz-DiosdAdo A, Balderas-López AJ, Ibáñez-Hernández M. 2019. Polyethylene glycol-coated gold nanoparticles as DNA and atorvastatin delivery systems and cytotoxicity evaluation. *J. Nanomater.* 2019, 5982047.

Zhang CY, Yeh HC, Kuroki MT, Wang TH. 2005. Single-quantum-dot-based DNA nanosensor. *Nat. Mater.* 4 (11), 826–831.

9 Biopolymers
Sources, Chemical Structures, and Applications

Cai Jen Sia, Lee Hoon Lim, and Anwar Usman

CONTENTS

9.1   INTRODUCTION

In the past few decades, the term "polymer" has been frequently used in various industries due to its immense characteristics in diverse applications. The polymer has a structure where any natural or synthetic substances are composed of multiple repeating sub-units by chemical bonding to form large molecules called macromolecules. There are various polymers used to manufacture different products and goods that are required in human daily life these days. Synthetic polymers have been recognized as the most used polymer in daily life. They are also known as man-made polymers that are derived from petroleum oil, which is manufactured in laboratories under a controlled environment. Additionally, they are commonly made up of carbon-carbon bonds as their backbone. In order to improve the chemical bonds of synthetic polymers that hold monomers together, a catalyst is added to speed up the reactions between monomers with the presence of both heat and pressure (Shrivastava, 2018).

Synthetic polymers had been acted as an important role in various material productions that are required in human daily life, agriculture, industry, and others. Hence, the abundant usage of these materials can lead to major environmental problems, such as waste plastics and water-soluble synthetic polymers in wastewater. The usage of synthetic polymers has widely increased annually, where approximately 140 million tons of synthetic polymers are being produced globally (Wankhade, 2020). In the biosphere, the degradation cycles of polymers are unlimited, as they are extremely stable (Kawalkar, 2014). It is noteworthy that the plastic and polymers are derived from petroleum-based polymers. There are various benefits of these petroleum-based polymers, such as their good strength, malleability, excellent processability, resistivity, chemical inertness, and physico-chemical properties. The use of these polymers is discouraged because they are non-biodegradable. As these polymers, in general, have been attached to be part of human daily life, and are stable and resistant to degradation, they can accumulate in the environments at the percentage of approximately 8% by weight and 20% by volume of the landfills (Premraj & Doble, 2005).

In order to minimize the risks to consumers, the environment, and the food chain, it is preferable to use industrial crops or non-food production for the manufacture of biomaterials (Ibrahim et al., 2019). Biopolymers seem to be a great preference for the replacement of synthetic polymers due to their eco-friendliness, biodegradability, safety for oral consumption, non-toxicity, and biocompatibility, which play an important role in various elementary research and applications. In fact, it is believed that biopolymers can replace synthetic polymers for about 30–90% in the future (Aaliya, Sunooj, & Lackner, 2021). Biopolymers are different polymeric biomolecules that can be formed within or derived from the cell of living organisms, such as plants, animals, and microorganisms, by complex metabolic processes or either from petroleum, which is recognized as the traditional source of polymers (Kawalkar, 2014; Christian, 2016). The structure of a biopolymer consists of two or more monomeric repeating units, which are covalently bonded to form larger or chain-like molecules (Shankar & Rhim, 2018). Additionally, the structures of biopolymers are simpler than those compared to synthetic polymers, so they can be easily distinguished. In biopolymers, there are complex molecules that gather together that support accuracy and define their 3D shapes and structures (Mohan et al., 2016).

Biopolymers are highly produced on a large scale for diverse applications. The usage of biopolymers has been increasing yearly even though they occupy a very small fraction of the polymer market. They are also recognized as biodegradable, which are well-matched in human body fluids or tissues; are easy to operate in the laboratory; and are capable of being ionized with conformational changes (Chow et al., 2008). They usually take several years to decompose in a natural environment, which can lead to environmental issues, such as water contamination, soil degradation, and the death of animals (Li et al., 2016). Most of the biopolymers are biodegradable but not all. The degradation capacity of biopolymers depends on various factors, such as polymer types, chemical composition, and environmental conditions (Aaliya, Sunooj, & Lackner, 2021). Biopolymers are degraded depending on the action of commonly occurring organisms and leave overdue organic by-products, such as carbon dioxide and water, which have no harmful effect on the environment. Biopolymers are only used in specific fields, depending on their cost, availability, moisture adsorption, thermal stability, mechanical behavior, degradation stability, and biocompatibility (Christian, 2016). Examples of biopolymers that can be chemically synthesized by humans from biological sources are vegetable oils, fats, sugars, resins, proteins, and amino acids (Deb et al., 2019).

In the environment, biopolymers offer the usage of renewable energy resources, which can reduce greenhouse gas emissions. They only required low energy to be synthesized and manufactured. In recent years, the usage of biopolymers had been increased in diverse fields, as documented by various papers that have been recently published (Ibrahim et al., 2019; Sivakanthan et al., 2020). Nowadays, various industries have been experiencing an extremely fascinating growth of biopolymers, which provide strong invention opportunities in many sectors, such as packaging, automotive, construction, sports, adhesives, and paints. The capacity of global biopolymer production is estimated to increase from 2.42 million tons in 2021 to about 7.59 million tons in 2026.

With the growth of interest in biopolymers and their fascinating properties as advanced materials for a wide range of applications, this chapter provides an overview of the classification, chemical structures, and applications of biopolymers. This chapter includes on a few sections, including the classification of biopolymers, structures of biopolymers, and applications of biopolymers, especially in food, food packaging, agriculture, wastewater treatment, cosmetic, biomedical, and high-technology applications.

9.2 CLASSIFICATION OF BIOPOLYMERS

Based on degradability, biopolymers can be divided into biodegradable and non-biodegradable groups, or either into bio-based and non-bio-based biopolymers, as shown in Figure 9.1. They may be classified as formed from natural and fossil fuels based on their source of origin. They can also be segregated into elastomers, thermoplastics, and thermosets due to their thermal condition response. According to their composition, biopolymers are divided into blends, laminates, and composite groups (George et al., 2020). Based on the polymer backbone, biopolymers are classified into polyamides, polysaccharides, polyesters, polycarbonates, and vinyl polymers. Biopolymers can also be classified as bioplastics, biosurfactant, biodetergent, bioadhesive, bioflocculant, and many more, depending on their applications.

In general, biopolymers are classified as natural biopolymers, chemically synthesized biopolymers, and microbial biopolymers based on their source of

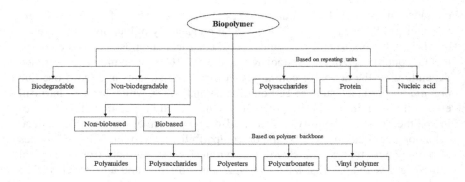

FIGURE 9.1 Different class of biopolymers that based on their biodegradability, polymer backbone, and type of monomers.

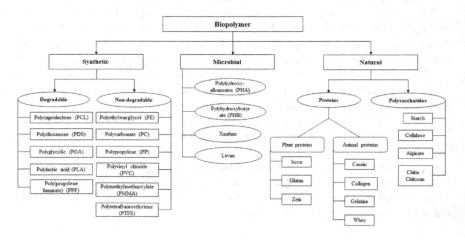

FIGURE 9.2 Classification of biopolymers with their examples.

raw materials, as shown in Figure 9.2. Natural biopolymers that are extracted from biomass can be found from the derivation of protein and polysaccharide-based biopolymers, which are produced from plants, animals, and microorganisms. Examples of protein-based biopolymers are albumin, collagen, keratins, zein, gelatin, sericin, fibrinogen, microfibrillar, wheat gluten, alginate, casein, soy protein, milk protein, egg protein, and whey protein (Gupta & Nayak, 2015; Awe et al., 2017; Nagarajan et al., 2019; Udayakumar et al., 2021a). In addition, examples of carbohydrate-based biopolymers are cellulose, agar, galactans, pectin, carrageenan, alginate, chitosan, gums, lignin, starch, hyaluronic acid, and alginic acid (Awe et al., 2017; Verma, Manjubala, & Narendrakumar, 2016;

Udayakumar et al., 2021b). Synthetic biopolymers that are chemically synthesized from biomass can be divided into degradable and non-degradable groups. The degradable group of synthetic biopolymers consists of polycaprolactone, polydioxanone, polyglycolic, polylactic acid, and poly(propylene fumarate), whereas the non-degradable group of synthetic biopolymers consists of polyethyleneglycol, polycarbonate, polypropylene, polyvinyl chloride, polymethylmethacrylate, and polytetrafluoroethylene. Lastly, microbial biopolymer consists of polyhydroxyalkanoates, poly(3-hydroxybutyrate-co-3-hydroxyvalerate), polyhydroxybutyrate, levan, curdlan, gellan, dextran, pullulan, and xanthan (Ibrahim et al., 2019; Udayakumar et al., 2021a).

9.3 SOURCES OF BIOPOLYMERS

As mentioned earlier, biopolymers can be derived from natural sources, such as plants, animals, microbes, and agricultural wastes. Plant sources of biopolymers are maize, rice, wheat, potatoes, sorghum, yams, cassava, banana, corn, tapioca, cotton, and barley, where biopolymers can be chemically synthesized from monomeric units, such as oils, sugars, and amino acids (Baranwal et al., 2022). The animal sources of biopolymers consist of cattle and marine sources, such as corals, sponges, fishes, lobsters, and shrimps (Udayakumar et al., 2021b). Microbial sources involve algae, fungi, and yeasts. Agro residues, crops, paper wastes, wood wastes, and green wastes are recognized as important materials that are carbohydrate-rich biomass-based sources. Vegetable oils attained from food producers are rich in triglycerides, such as soybean, corn, sunflower, castor, safflower, rapeseed, jojoba, teak, tung, linseed, castor, fish, and meadowfoam oils, which act as great replacements for the combination of natural polymers (Baranwal et al., 2022). There are various biopolymers being found. Some will be discussed subsequently.

9.3.1 CARRAGEENAN

Carrageenan is the major polysaccharides component obtained by extraction from the extracellular matrix of certain red seaweeds (*Rhodophyta*), such as *Eucheuma, Gigartina, Chondrus*, and *Hypnea* species (Guo et al., 2022). The word "carrageenan" comes from an informal Irish name of seaweed named "carrageen," which means "little rock" (Campo et al., 2009). It was first discovered in Ireland in 1810. The chemical structure of carrageenan is formed by alternate units of β-d-galactose and 3,6-anhydro-α-d-galactose that are linked by α-(1,3) and β-(1,4) glycosidic linkage. Carrageenan can be classified into λ, κ, ι, ε, and μ, which all contain up to 35% of the sulfate groups. Figure 9.3 shows the chemical structures of the most common carrageenan.

The viscosity of carrageenan depends on the concentration, temperature, existence of additional solutes, type of carrageenan, and its molecular weight. The classification of carrageenan depends on its solubility in potassium

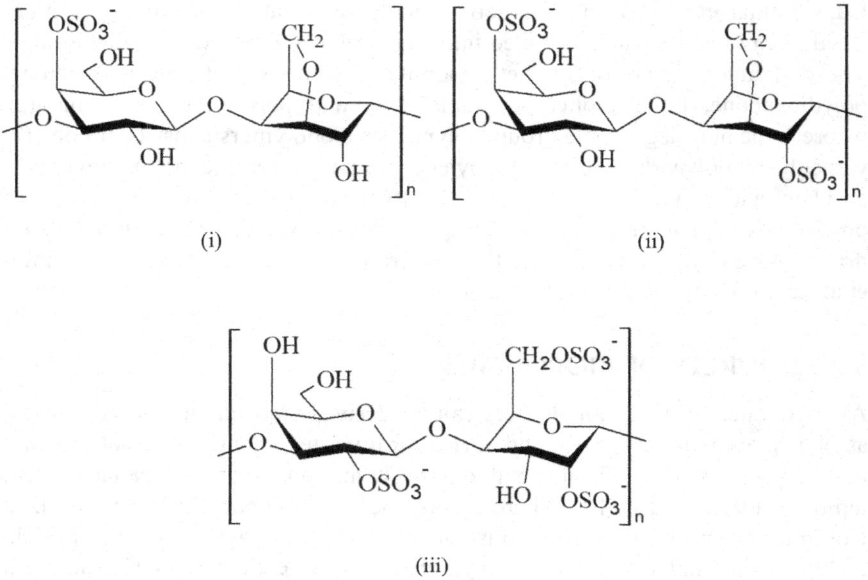

FIGURE 9.3 The chemical structures of (i) κ-carrageenan, (ii) ι-carrageenan, and (iii) λ-carrageenan.

chloride. λ-carrageenans are easily soluble in both cold or hot aqueous solutions, whereas κ-carrageenans are only soluble in a hot solution. κ-carrageenans and ι-carrageenans form gels in the presence of potassium or calcium ions, where λ-carrageenans do not form gels (Michel, Mestdagh, & Axelos, 1997). Although carrageenans have no nutritional value, it is widely used for food preparation due to its excellent solubility and texture, acting as a gelling agent, thickener, stabilizer, and emulsifier. The concentration of carrageenan used in food products is approximately about 0.005–2.0% by weight (Necas & Bartosikova, 2013). It has been widely used for milk products, processed meats, dietetic formulations, infant formula, toothpaste, cosmetics, skin preparations, pesticides, and laxatives. Campo et al. highlighted that about 70–80% of total global production in the food industry, which is about 45,000 tonnes produced annually, comes from dairy products (45%) and meat and meat by-products (30%) (Campo et al., 2009). Hence, the total carrageenan market is estimated to be about $300 million annually (Campo et al., 2009).

9.3.2 CELLULOSE

In 1838, cellulose was first discovered by Payen (Liu et al., 2021). Cellulose is the most abundant natural biopolymer that has a long chain of linked glucose molecules arranged in linear form. It has a chemical formula of $(C_6H_{10}O_5)_x$, which

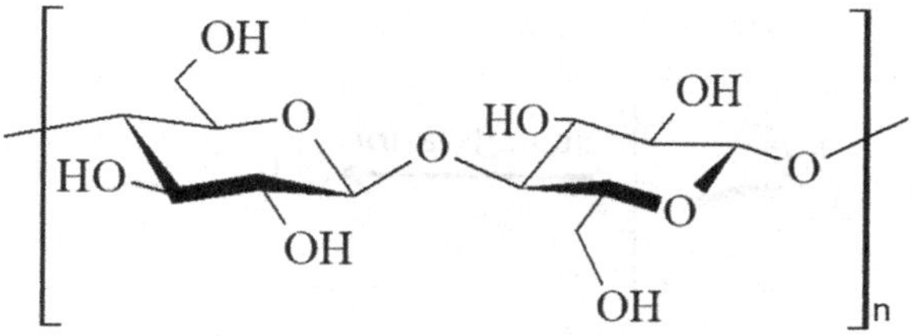

FIGURE 9.4 The chemical structure of cellulose.

contains thousands of $\beta(1 \rightarrow 4)$-linked glucose chains with hydrogen bonds that are formed between both hydroxyl groups and oxygen atoms inside a glucose chain and also among the nearest chains (see Figure 9.4). Additionally, the chemical formula of cellulose will be depending on different aspects, such as types of plants, growth environment, and maturity. Cellulose is considered an infinite source of raw material that is of high interest for environment-friendly and biocompatible products due to its strong hydrogen bonding and insolubility in solvents. Hence, it is chemically altered into the derivation of ether, ester, and acetal.

Approximately 7.5×10^{10} tons of cellulose are produced globally every year (Riswati et al., 2021). The sources of cellulose are flax, hemp, jute, wood, and cotton. It is mostly used in the manufacture of paper, cardboard, textiles, fiber, and biofuel industries (Venkateshaiah et al., 2020). The outstanding strength of wood comes from cellulose where there are long chains of sugar molecules linked together. Cellulose plays as the main component in the cell walls of plants, which have the basic building block for textiles and paper. The most common cellulose without undergoing any process is cotton, which is used in numerous textile applications and membranes.

9.3.3 CHITIN AND CHITOSAN

Chitin and chitosan were first discovered in the year 1894. Chitin is widely distributed in nature and is recognized as the second most abundant biopolymer among all the polysaccharides after cellulose. It can be found in the major structural component that acts as supporting tissues in the exoskeletons of crustaceans as well as the cell wall of particular fungi and algae (Rinaudo, 2006; Mohammed, Williams, & Tverezovskaya, 2013; Arbia et al., 2013). It has a chemical structure of poly-β-$(1 \rightarrow 4)$-N-acetyl-D-glucosamine, as shown in Figure 9.5.

Chitin is hard, unbendable, and highly insoluble in aqueous media, which can replicate a large number of hydrogen bonds between acetamido groups intra- and inter-polymer chains (Schmitz et al., 2019). Additionally, it is an extremely insoluble substance because of the low solubility of cellulose and chemical non-reactivity (Anitha et al., 2014). In general, the biopolymer can be synthesized by a large number of living organisms that can be used to estimate the amount

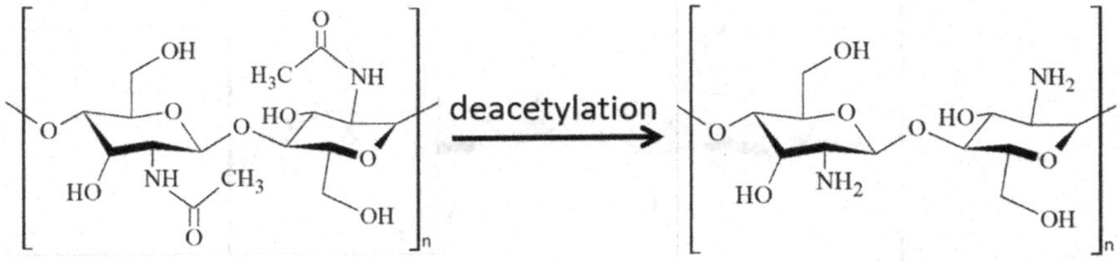

FIGURE 9.5   The chemical structures of chitin and chitosan.

of chitin (Rinaudo, 2006; Schmitz et al., 2019). The estimation for chitin production annually can increase to 10^{12} tonnes (Varun et al., 2017). The seafood industry usually produces about one million tonnes of chitin as waste, followed by converting the chitin into low-value-added products, such as fertilizers and animal foods (Schmitz et al., 2019). The main source of chitin is commonly only crab and shrimp shells. Crustacean shells commonly contain about 30–40% of protein, 30–50% of calcium carbonate, and 20–30% of chitin (Kumari & Rath, 2014; Rashmi et al., 2016).

Demineralization and deproteinization are two major processes that needed to be done in order to extract chitin from crustacean shells. Chitin will then undergo deacetylation treatment, where the presence of acetyl groups is removed in order to form chitosan. As mentioned earlier, chitin is highly insoluble in aqueous media, but it is soluble under acidic conditions when deacetylated to chitosan (Mohammed, Williams, & Tverezovskaya, 2013). Chitosan has the chemical structure of a linear amino polysaccharide, which links β-(1-4)-linked D-glucosamine and N-acetyl-D-glucosamine (Divya, Rebello, & Jisha, 2014).

Chitosan is found to be highly insoluble due to its solubility and is low in chemical reactivity, but it is soluble in acidic aqueous solutions (Schmitz et al., 2019). Chitosan and its derivatives are found to be renewable and non-toxic compounds, which consist of various biological activities; for example, anti-cancer, antioxidant, anti-microbial, anticoagulant, antihypertensive, antidiabetic, anti-obesity, antiallergic, anti-inflammatory, neuroprotective, and matrix metalloproteinases inhibitory agents (El-Aidie, 2018; Salah et al., 2013; Yen, Yang, & Mau, 2008; Goy, De Britto, & Assis, 2009; Vongchan et al., 2003). It has been recognized as a functional biopolymer due to its physicochemical and biological properties that have been widely used in various applications, such as in the area of nutrition, cultivation, medication, packaging additive, food, textile, dietary supplement, wastewater treatment, agricultural industry, preservative, and pharmaceutical and cosmetic industry (Schmitz et al., 2019; Santos et al., 2020; Muñoz et al., 2018). The presence of chitosan in industries was first reported in Japan and the United States in 1970; they were considered the world leaders in the production of chitosan that is economically attractive and profitable (Santos et al., 2020). Additionally, there are approximately 15 industries in Japan manufacturing chitin and chitosan commercially in the year 1986 (Santos et al., 2020).

Chitosan was being used in the healthcare industry (65%), followed by the agricultural industry (12%), wastewater treatments (7%), food and beverage industry (6%), and immobilization and biotechnology (5%) (El-Aidie, 2018; Li et al., 1992). Researchers have conducted a study on the usage of chitosan for the transportation of medicine, specifically for colon disease treatment, and also its usage as a dietary supplement for decreasing cholesterol and regulatory body weight (Kumar, Gowda, & Tharanathan, 2004). It is worth noting that the United States Food and Drug Administration (USFDA) has acknowledged that chitosan can be used as a food preservative from a natural origin, as a Generally Recognized as Safe (GARS) food additive (El-Aidie, 2018).

9.3.4 COLLAGENS

Collagens are the most abundant proteins in the human body, which contain a total protein mass of approximately 30% (Ricard-Blum, 2011). It was discovered by Payen in the year 1838 (Yadav et al., 2015). In human health, collagen has been playing an important role, where the breakdown and reduction of natural collagen in the body are related to several health problems (Mohan et al., 2016). Exogenous collagen is designed for usage in medical and cosmetic fields, which helps with the healing and repairing of body tissues. In the 1950s, type I collagen was discovered by Ramachandran (Venkateshaiah et al., 2020). It is the most common that consists of three polypeptide subunits, where amino acids are linked to arginine, glycine, proline, hydroxyproline, and lysine, as shown in Figure 9.6 (Venkateshaiah et al., 2020). Type I collagen can be found in the

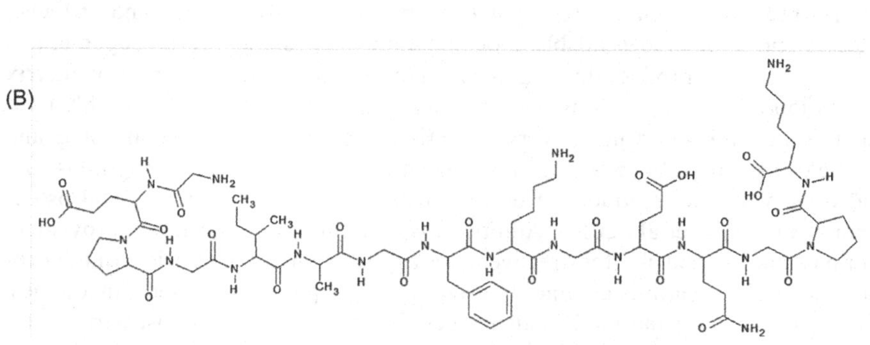

(A)

(B)

FIGURE 9.6 The chemical structures of (A) type I collagen and (B) type II collagen.

skin, tendons, ligaments, and bones, whereas type II collagen can be found in the cartilage.

Type II collagen was discovered by Miller and Matukas in 1969, and 26 new types of collagens had being introduced according to their molecular biology and gene cloning (Ricard-Blum, 2011). Hence, the molecular structure of type II collagen is similar to type I collagen, but in type II collagen, the chain is homotrimer linked with $[\alpha1(II)]_3$, and the major triple helical area with 1,014 residues of the repeated -Gly-Xaa-Yaa- sequence is edged by N- and C-propeptides, as in type I collagen (see Figure 9.6B) (Bächinger et al., 2010).

The sources of collagen are fish and poultry waste (Venkateshaiah et al., 2020). Collagen is commonly used in various medical applications, such as in tissue infection treatment, pharmaceutical products, drug delivery systems, gene therapy, and edible casings (Sionkowska et al., 2017). In medical products, the derivations of collagen are from human, bovine, porcine, and ovine sources. The controllable aspects that harm collagen production are sunlight, smoking, and high sugar consumption (Mohan et al., 2016).

9.3.5 GELATIN

Gelatin is a protein that is derived from animal products by partial hydrolysis of collagen that is obtained from the skin, bones, and tissues of animals. Depending on the method used for its production, gelatin can be classified into Type-A and Type-B. Type-A gelatin is obtained by acid hydrolysis of collagen, which consists of 18.5% of nitrogen and an isoelectric point (pI) ranging from 7.0 to 9.0 (Mohan et al., 2016; Kommareddy, Shenoy, & Amiji, 2007). However, Type-B gelatin is obtained by alkaline hydrolysis of collagen, which consists of 18% of nitrogen with the absence of amide groups and a pI value ranging from 4.8 to 5.0 (Mohan et al., 2016; Kommareddy, Shenoy, & Amiji, 2007). Gelatin is a diverse combination of single- or multi-stranded polypeptides, each with prolonged left-handed proline helix forms, which contain between 300 and 4,000 amino acids. Figure 9.7 shows the chemical structures of gelatin. Gelatin can be modified with the combination of biomolecules and nanoparticles for numerous applications due to the presence of various functional groups, such as $-NH_2$, $-SH$, and $-COOH$. It is found to be biodegradable, biocompatible, hemostatic, pro-angiogenic, and cross-linked to produce hydrogels and non-immunogenic substrate of matrix metalloproteinases (Dash, Foston, & Ragauskas, 2013). Hence, it is highly used in the application of drug delivery, bone tissue engineering, wound dressing, and gene transfection. In the human body, gelatin is not just used for weight loss but also used for the treatment of osteoarthritis, rheumatoid, arthritis, and osteoporosis (Mohan et al., 2016). Additionally, gelatin helps with the improvement of hair quality and shortens recovery for any sports-related injuries. In the production field, gelatin is commonly used for the preparation of foods, medicines, and cosmetics. In the food industry, gelation creations use almost particularly water or aqueous polyhydric alcohols as solvents for candy, marshmallows, or

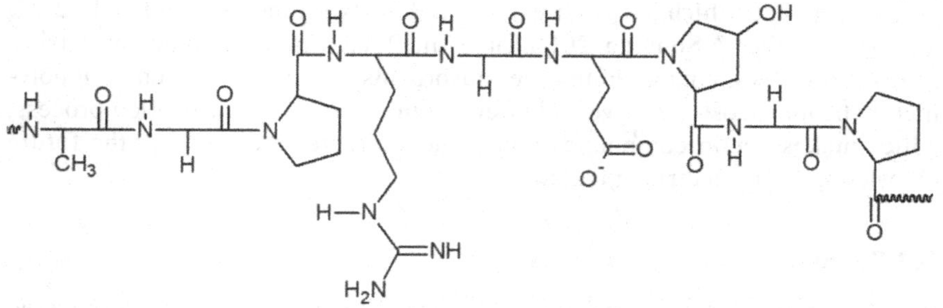

FIGURE 9.7 The chemical structure of gelatin.

preparations of dessert. The protective colloid property of gelatin helps to prevent the crystallization of ice and sugar in human dairy products and frozen food products (Keenan, 2012).

9.3.6 PECTIN

Pectin is an anionic biopolymer that is recognized as one of the important polymers that are commonly used in the food industry. It is soluble in water, biocompatible, and non-toxic. It has a linear polysaccharide with a high molecular weight that can be extracted from the cell walls of plants. The chemical formula of pectin is $C_6H_{10}O_7$, which has a molecular mass of 194.14 g/mol. The structure of pectin is designed from D-galacturonate molecules linked together in chains by α-(1–4) glycosidic linkages, where carboxyl or hydroxyl groups are extended along the backbone, as shown in Figure 9.8. In the food and beverage industry, pectin is used as a thickening agent, colloidal stabilizer, texturizer, and emulsifier in foods, and coating on fruits or vegetables (Mărtău, Mihai, & Vodnar, 2019). In medical industries, pectin is commonly used for

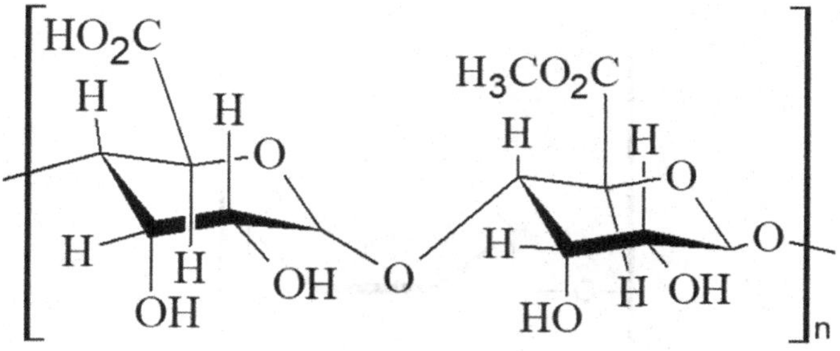

FIGURE 9.8 The chemical structure of pectin.

colon treatment, which has been widely studied over time (Sriamornsak, 2011; Wong, Colombo, & Sonvico, 2011; Liu et al., 2003; Munarin, Tanzi, & Petrini, 2012). Nowadays, compared to other adsorbents, the multifunctional biopolymer is the most cost-effective and effective in terms of the adsorption process, which makes pectin compounds a possible alternate adsorbent for the future (Moslemi, 2021; Shahrin et al., 2021).

9.3.7 STARCH

Starch is the second main agricultural product after cellulose. It was discovered by Leeuwenhoek in 1716 (Seetharaman & Bertoft, 2012). It is recognized as a low-cost biodegradable biopolymer and the most abundant storage polysaccharide in plants. The chemical formula of starch molecules is $(C_6H_{10}O_5)_x$, where the polysaccharide consists of glucose monomers combined in α-1,4-linkages (see Figure 9.9). According to the Agriculture and Food Development Authority of the United States, the worldwide annual production of starch is about 90 million tons. Starch with proper treatment can be used to make many commercial goods due to its sensitivity to moisture and poor mechanical properties (Mohan et al., 2016). It can be derived from various agricultural raw materials, such as potatoes, wheat, tapioca, rice, and corn, which can normally be found in roots, stalks, and crop seeds in plants (Ibrahim et al., 2019). In plants, starch mainly consists of amylose (30%), amylopectin (70%), and fats and proteins (less than 1%) (Ibrahim et al., 2019). Plants need starch as their energy storage, humans need it as part of their diet, and industries need it for numerous commercial goods, such as in the production of paper, textiles, and adhesives. In addition, starch is commonly used for the production of syrups that accommodate glucose, fructose, or maltose, which are broadly used in industries (Goyal, Gupta, & Soni, 2005). Hence, the importance of starch in plants and humans is undeniable. Due to its biodegradable and renewable nature, starch has been used as synthetic additives in various products that involved plastics, detergents, pharmaceutical tablets, pesticides, cosmetics, and oil-drilling fluids (Mohan et al., 2016).

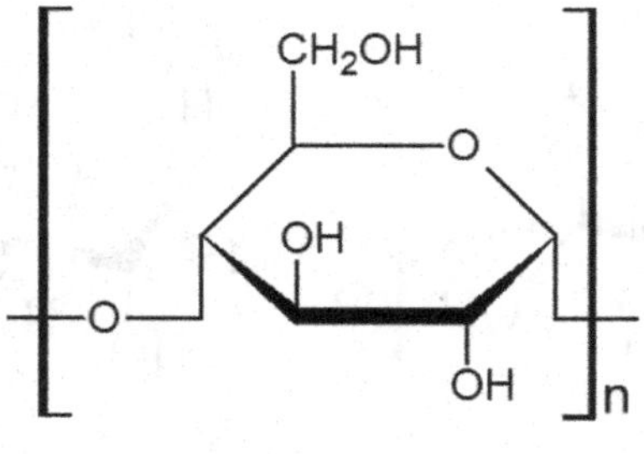

FIGURE 9.9 The chemical structure of starch.

9.4 APPLICATIONS OF BIOPOLYMERS

As mentioned, biopolymers are used to expand the acts of active molecules in a product due to their biocompatible and biodegradable nature. There are various applications of biopolymers that have been discovered that are widely used for daily life. In fact, numerous industrial applications commonly use biopolymers in food packaging, agriculture, water purification, wastewater treatment, cosmetics, biomedical applications, and many more.

9.4.1 FOOD INDUSTRY

Biopolymers play an important role in various food industries. They are widely used in the production of food packaging materials, food coatings, and encapsulation matrices for functional foods. They are safe for oral consumption and consist of special solutions that help to improve product shelf life. At the same time, biopolymers contribute to minimizing the overall carbon footprint related to food packaging (Ibrahim et al., 2019). The most common biopolymers used in food packaging are polylactic acid, polyhydroxyalkanoates, cellulose, chitosan, starch, agar that are derived from carbohydrates, whey protein, collagen that are derived from proteins, and many more, which can be manufactured by conventional equipment (Ibrahim et al., 2019; Shankar & Rhim, 2018; Mohan et al., 2016). However, these materials have been broadly used in various monolayer and multilayer applications in the food-packaging area. In fact, starch and PLA biopolymers are found to be the most commonly used types of biodegradable materials due to their sustainable properties and wide commercial availability.

For bio-based food packaging applications, the main concern is their barrier properties. Protein and carbohydrate packaging films are both considered to be good barriers against oxygen from a lower level to an intermediate qualified humidity, and have good mechanical properties. Nevertheless, the barrier has poor moisture resistance when the water vapor transmits through the packaging due to its hydrophilic nature, which can affect food quality. As a result, this causes shelf life to be shorter, costs to increase, and ultimately, more waste to be created. These barrier properties of biopolymers can be overcome by adding numerous nanofillers, such as nanoclays and metal oxide nanoparticles (Mohan et al., 2016). The main objectives of food packaging are the protection and safety of food from physical, chemical, or biological damages from the manufacture time until expiration. Furthermore, the replacement of petroleum-based packaging materials with bio-based films and containers provides good advantages due to its sustainable and greener image as well as improvement in the technical properties.

Biopolymers also show great potential for encapsulation purposes. Encapsulation is a useful technique that is used to enhance the carriage of bioactive molecules and living cells to foods by entrapping active agents within a carrier material. This process has been applied to prolong the shelf life of active food mechanisms, ensure their stability, achieve their controlled release, and enhance their functional properties (Gürbüz et al., 2020). The entrapped material is also known as a core, fill, internal phase, payload, or active material, which can be in the form of a liquid, solid, or gas. Furthermore, the material used to coat the active agents is known as a coating material, wall, capsule, membrane, shell, matrix, or carrier material (Timilsena, Haque, & Adhikari, 2020). The most common materials used to make these protective shells of encapsulates must have the properties of being food-grade, biodegradable, and capable of forming a barrier between its internal phase and surroundings (Nedovic et al., 2011). Biopolymers that are suitable to be used for encapsulation in food applications are polysaccharides, proteins, and lipids. Encapsulation technology is invented to entrap the active materials in small capsules, which can release their contents at controlled rates after a long time and under specific circumstances. This technology has increased great interest in the pharmaceutical and food industries. In the pharmaceutical industry, encapsulation is mainly applied for drug and vaccine delivery. Meanwhile, in the food industry, it acts as an addition to functional compounds in food products. The usage of functional compounds is to control the flavor, color, texture, or preservation properties of food products. In addition, encapsulation also helps to improve the stability of bioactive compounds during processing and storage, and to avoid unwanted interactions with food matrices (Nedovic et al., 2011). Encapsulation of food compounds can be applied in numerous techniques, such as spray drying, spray-bed drying, fluid-bed coating, spray chilling, spray cooling, or melt injection (Nedovic et al., 2011; Gibbs et al., 1999).

Biopolymers are utilized to prepare edible packaging films for food products. In this sense, polysaccharides and proteins are commonly used to manufacture packaging films, and they are safe for oral consumption. These edible films have the potential to replace synthetic materials, which helps to decrease packaging waste and diminish environmental pollution. Lastly, biopolymers can act as emulsifying, thickening, and moisture-retaining agents for the purpose of improving the stability and physicochemical properties of food emulsions (Gheorghita et al., 2021).

9.4.2 BIOMEDICAL APPLICATIONS

Biopolymers have been recognized as the most suitable materials for developing various biomedical applications because they can naturally degenerate in the human body without forming any harmful side effects (Ibrahim et al., 2019). Biopolymers play a significant role in maintaining and promoting health by decreasing the risk of degenerative diseases that are caused by free radicals and lowering the risk of pathogenic microorganisms either as antioxidants or antimicrobials (Sivakanthan et al., 2020). They are generally used in different biomedical applications, such as tissue engineering, pharmaceutical carriers, and medical devices. In the pharmaceutical field, biopolymers are used to manufacture the packaging for pharmaceutical products, where they help to protect the pills, nutraceuticals, drugs, blood and blood products, surgical devices, powders, liquid, and solid and semisolid dosage forms (Kasar, Tribhuwan, & Khode, 2020). The pharmaceutical packaging has to cause no harm to the patient's condition and health, and be environmentally friendly (Zadbuke et al., 2013).

For instance, gelatin is the most commonly used biopolymer in the biomedical industry due to its low manufacturing costs, low manufacturing complexity, and excellent dissolution rates of active pharmaceutical ingredients, which are guaranteed. It also helps in protecting sensitive ingredients from oxygen, light, microbial growth, and other contamination. It is used for wound dressing, tissue engineering, stem cell therapy, adhesive, and many more (Ibrahim et al., 2019; Pereira et al., 1998; Rodrigues & Emeje, 2012; Pal, Banthia, & Majumdar, 2006). In addition, simple porogens, such as solvents and gases, are required for the fabrication of porous gelatin scaffolds and films. Hence, this technique creates scaffolds to carry the drugs or nutrients to the wound area for healing (Ibrahim et al., 2019). Collagen is another bio-based polymer that plays an important role in tissue engineering applications due to its ability to repair bone, tendon, ligament, and vascular and connective tissues. Other than tissue engineering applications, it is also applied to ophthalmic formulations, collagen shields, eye implants, ocular drug delivery, drug delivery, gene delivery, protein delivery, and many more (Rathore & Nema, 2009; Eshar, Wyre, & Schoster, 2011; Liu et al., 2006; Le Bourlais et al., 1998; Ruszczak & Friess, 2003; Capito & Spector, 2007; Chan, So, & Chan, 2008; Jacob & Gopi, 2021).

9.4.3 WATER PURIFICATION AND WASTEWATER TREATMENT

Water can be polluted in various ways due to city sewage and industrial waste discharge. Water pollution usually happens when the water supply is contaminated with soils or groundwater systems, and from the atmosphere by rainwater. Soils and groundwater consist of residue that comes from agricultural chemical fertilizers and improper disposal of industrial wastes. Water plays an important role in daily life and is vital for human health, food production, ecosystem function, and financial increase. Therefore, nanotechnology has been developed for safe drinking water through a successful purifying mechanism (Mohan et al., 2016). The development of nanomaterials that contain antibacterial and antifungal materials for water purification and wastewater treatment is considered the most effective way to provide clean and safe drinking water for humans. In addition, many biopolymer-based nanocomposites can eliminate all toxic heavy metals, such as lead and arsenic, from contaminated water with the use of antibacterial agents. For example, chitosan is one of the biopolymers that has been used as flocculants in water treatment processes and can be biodegraded in the environment over a long period.

9.4.4 COSMETICS INDUSTRY

In this modern era, cosmetic industries have been increasing rapidly, and the global market for cosmetics will reach about $429.8 billion by 2022 (Gupta et al.,

2022). The production of cosmetics commonly requires the grouping of pH, color, stabilizers, preservatives, water, fragrance, thickener, and emulsifiers. In fact, some cosmetic products are also required to have pharmaceutical effects; hence, active formulations that are so-called cosmeceuticals are invented. Cosmeceuticals help to improve the appearance of healthy skin by transporting essential nutrients. Cosmetic industries are where various biopolymers are being used due to their low-cost, strength, and flexibility. Biopolymers that are commonly used in the cosmetic industry are chitosan, cellulose, collagen, keratin, gum, starch, polyhydroxyalkanoates, and many more. For instance, starch can form naturally as small white granules that differ in terms of size and shape, depending on their origin. In cosmetic applications, starch functions as adsorbent of fats, enhances skin texture, and forms excellent powders. Guar gum is another biopolymer that is broadly used to produce face masks, lipsticks, shampoos, conditioners, and many more. It has water solubility properties and non-allergic properties for use in all types of skin.

9.4.5 MODERN TECHNOLOGIES IN BIOPOLYMERS

Both natural and synthetic biopolymers are recognized as essential advanced materials in modern technologies. Many industries have been utilizing biopolymers in the development of various modern technologies. For instance, contact lenses are one of the most popular products that have been widely used as a convenient alternative to spectacles by more than 150 million people in the world. In 2015, the estimate of its global market size is worth about $7.1 billion (Musgrave & Fang, 2019). The main aims of contact lens applications are to help correct human vision and use it in cosmetics for therapeutic reasons. A suitable polymeric material is highly recommended in order to manufacture contact lenses. The production of contact lenses is dependent on their lens thickness, wettability, O_2 permeability, H_2O content, and mechanical properties. Additionally, the biomaterials for manufacturing soft contact lenses have to be optically transparent, chemically stable, and thermally stable, and be able to provide high tensile and tear strength for durability of lens handling (Goda & Ishihara, 2006). The most common monomers and polymers used in manufacturing contact lenses are polymethyl methacrylate, polyvinyl alcohol, polyethylene glycol, dimethyl methacrylate, hydroxyethyl methacrylate, N-vinyl pyrrolidone, ethylene glycol dimethylacrylate, polydimethylsiloxane, and 3-[tris(trimethylsiloxy)silyl]propyl methacrylate. Hydrogels have been commonly used to manufacture soft contact lenses since the 1980s. Hydrogel contact lenses are extremely thin, malleable, affordable, and comfortable, and adhere to the surface of the eyes. The main hydrogel material used for soft contact lenses is hydroxyethyl methacrylate. The oxygen permeability of these hydrogel contact lenses is directly proportional to the water content. Hydrogels can be derived from both natural and synthetic biopolymers. Sources of hydrogels are collagen, silk fibroin, hyaluronic acid, chitosan, alginate, and many more.

Cigarettes have been a widely known tobacco product that is harmful to human health. When burning, cigarettes can create over 7,000 chemicals, where at least 69 of the chemicals are toxic. According to WHO, about 22.3% of the global population uses tobacco. Smoking is not just harmful to smokers but also to passive smokers. Hence, a good filter in cigarettes has been investigated in order to minimize the risk of smokers. Cellulose acetate has been widely used for cigarette filters due to its eco-friendliness and availability. These filters help with the removal of tar and nicotine while keeping a favorable taste to smokers. The filters only assist in blocking major tar particles but allow minor tar to travel deeper into human lungs. Cellulose acetate can be prepared from wood pulp through reactions with acetic acid, acetic anhydride, and followed by sulfuric acid to produce cellulose triacetate.

9.5 CONCLUSION

In summary, biopolymers have been a popular research subject since a few decades ago. Biopolymers are ubiquitous in plants, animals, and microorganisms, and can be extracted and isolated through a few steps of chemical reactions. The important characteristic of biopolymers is their non-toxicity, and most of them are biocompatible and biodegradable, so they can be applied in many fields, such as food, food packaging, agriculture, wastewater treatment, cosmetics, biomedicine, and high technology. Biopolymers have different backbones, which can be classified into polyamides, polysaccharides, polyesters, polycarbonates, and vinyl polymers. Moreover, biopolymers consist of several active functional groups, such as amide, amino, acetyl, hydroxyl, and sulfonate groups. These functional groups interconnect the biopolymer chains through intra- and inter-strand hydrogen bonding interactions and stabilize the three-dimensional structures of the biopolymers. Biopolymers can be chemically modified in order to enhance their physical and chemical properties for desired applications. Therefore, further development of designing, exploring, and synthesizing new derivatives of biopolymers in the future would provide more fascinating advanced materials for such applications.

REFERENCES

Aaliya B, Sunooj KV, Lackner M. 2021. Biopolymer composites: A review. *Int. J. Biobased Plast.* 3 (1), 40–84. https://doi.org/10.1080/24759651.2021.1881214.

Anitha A, Sowmya S, Kumar PTS, Deepthi S, Chennazhi KP, Ehrlich H, Tsurkan M, Jayakumar R. 2014. Chitin and chitosan in selected biomedical applications. *Prog. Polym. Sci.* 39 (9), 1644–1667. https://doi.org/10.1016/j.progpolymsci.2014.02.008.

Arbia W, Arbia L, Adour L, Amrane A. 2013. Chitin extraction using biological methods. *Food Technol. Biotechnol.* 51 (1), 12–25.

Awe OW, Zhao Y, Nzihou A, Minh DP, Lyczko N. 2017. A review of biogas utilisation, purification and upgrading technologies. *Waste Biomass Valorization* 8 (2), 267–283. https://doi.org/10.1007/s12649-016-9826-4.

Bächinger HP, Mizuno K, Vranka JA, Boudko SP. 2010. Collagen formation and structure. *Compr. Nat. Prod. II Chem. Biol.* 5, 469–530. https://doi.org/10.1016/b978-008045382-8.00698-5.

Baranwal J, Barse B, Fais A, Delogu GL, Kumar A. 2022. Biopolymer: A sustainable material for food and medical applications. *Polymers* 14 (5), 1–22. https://doi.org/10.3390/polym14050983.

Campo VL, Kawano DF, da Silva DB, Carvalho I. 2009. Carrageenans: Biological properties, chemical modifications and structural analysis—A review. *Carbohydr. Polym.* 77 (2), 167–180. https://doi.org/10.1016/j.carbpol.2009.01.020.

Capito RM, Spector M. 2007. Collagen scaffolds for nonviral IGF-1 gene delivery in articular cartilage tissue engineering. *Gene Ther.* 14 (9), 721–732. https://doi.org/10.1038/sj.gt.3302918.

Chan OCM, So KF, Chan BP. 2008. Fabrication of nano-fibrous collagen microspheres for protein delivery and effects of photochemical crosslinking on release kinetics. *J. Control. Release* 129 (2), 135–143. https://doi.org/10.1016/j.jconrel.2008.04.011.

Chow D, Nunalee ML, Dong Woo Lim DW, Simnick AJ, Ashutosh Chilkoti A. 2008. Peptide-based biopolymers in biomedicine and biotechnology. *Mater. Sci. Eng. R Rep.* 62 (4), 125–155. https://doi.org/10.1016/j.mser.2008.04.004.

Christian SJ. 2016. Natural fibre-reinforced noncementitious composites (biocomposites). In: *Nonconventional and Vernacular Construction Materials-Characterisation, Properties and Applications*, Harris KA, Sharma B. (Eds.), Elsevier, Amsterdam, Netherlands, pp. 169–187. https://www.sciencedirect.com/science/article/pii/B9780081000380000056?via%3Dihub.https://doi.org/10.1016/b978-0-08-100038-0.00005-6.

Dash R, Foston M, Ragauskas AJ. 2013. Improving the mechanical and thermal properties of gelatin hydrogels cross-linked by cellulose nanowhiskers. *Carbohydr. Polym.* 91 (2), 638–645. https://doi.org/10.1016/j.carbpol.2012.08.080.

Deb PK, Kokaz SF, Abed SN, Paradkar A, Tekade RK. 2019. Pharmaceutical and biomedical applications of polymers. In: *Basic Fundamental of Drug Delivery*, Tekade RK (Ed.), Elsevier, Amsterdam, Netherlands, pp. 203–267. https://doi.org/10.1016/B978-0-12-817909-3.00006-6. https://www.sciencedirect.com/science/article/pii/B9780128179093000066.

Divya K, Rebello S, Jisha MS. 2014. A simple and effective method for extraction of high purity chitosan from shrimp shell waste. *Int. Conf. Adv. Appl. Sci. Environ. Eng.*, 141–145.

El-Aidie SAAM. 2018. A review on chitosan: Ecofriendly multiple potential applications in the food industry. *Int. J. Adv. Life Sci. Res.* 1 (1), 1–14. https://doi.org/10.31632/ijalsr.2018v01i01.001.

Eshar D, Wyre NR, Schoster JV. 2011. Use of collagen shields for treatment of chronic bilateral corneal ulcers in a pet rabbit. *J. Small Anim. Pract.* 52 (7), 380–383. https://doi.org/10.1111/j.1748-5827.2011.01077.x.

George A, Sanjay MR, Srisuk R, Parameswaranpillai J, Siengchin S. 2020. A comprehensive review on chemical properties and applications of biopolymers and their composites. *Int. J. Biol. Macromol.* 154, 329–338. https://doi.org/10.1016/j.ijbiomac.2020.03.120.

Gheorghita R, Anchidin-Norocel L, Filip R, Dimian M, Covasa M. 2021. Applications of biopolymers for drugs and probiotics delivery. *Polymers* 13 (16), 2729. https://doi.org/10.3390/polym13162729.

Gibbs BF, Kermasha S, Alli I, Mulligan CN. 1999. Encapsulation in the food industry: A review. *Int. J. Food Sci. Nutr.* 50 (3), 213–224. https://doi.org/10.1080/096374899101256.

Goda T, Ishihara K. 2006. Soft contact lens biomaterials from bioinspired phospholipid polymers. *Expert Rev. Med. Devices* 3 (2), 167–174. https://doi.org/10.1586/17434440.3.2.167.

Goy RC, De Britto D, Assis OBG. 2009. A review of the antimicrobial activity of chitosan. *Polimeros* 19 (3), 241–247. https://doi.org/10.1590/S0104-14282009000300013.

Goyal N, Gupta JK, Soni SK. 2005. A novel raw starch digesting thermostable α-amylase from *Bacillus sp.* I-3 and its use in the direct hydrolysis of raw potato starch. *Enzyme Microbial Technol.* 37 (7), 723–734. https://doi.org/10.1016/j.enzmictec.2005.04.017.

Guo Z, Wei Y, Zhang Y, Xu Y, Zheng L, Zhu B, Yao Z. 2022. Carrageenan oligosaccharides: A comprehensive review of preparation, isolation, purification, structure, biological activities and applications. *Algal Res.* 61, 102593. https://doi.org/10.1016/j.algal.2021.102593.

Gupta P, Nayak KK. 2015. Characteristics of protein-based biopolymer and its application. *Polym. Eng. Sci.* 55 (3), 485–498. https://doi.org/10.1002/pen.23928.

Gupta S, Sharma S, Nadda AK, Husain, MSB, Gupta A. 2022. Biopolymers from waste biomass and its applications in the cosmetic industry: A review. *Mater. Today Proc.* 68 (4), 872–879. https://doi.org/10.1016/j.matpr.2022.06.422.

Gürbüz E, Keresteci B, Günneç C, Baysal G. 2020. Encapsulation applications and production techniques in the food industry. *J. Nutr. Heal. Sci.* 7 (1), 106.

Ibrahim S, Riahi O, Said SM, Sabri MFM, Rozali S. 2019. Biopolymers from crop plants. *Ref. Modul. Mater. Sci. Mater. Eng.*, 1–10. https://doi.org/10.1016/b978-0-12-803581-8.11573-5.

Jacob J, Gopi S. 2021. Isolation and physicochemical characterization of biopolymers. In: *Biopolymers and Their Industrial Applications*, Thomas S, Gopi S, Amalraj A. (Eds.), Elsevier, Amsterdam, Netherlands, pp. 45–79. https://doi.org/10.1016/b978-0-12-819240-5.00003-1. https://www.sciencedirect.com/science/article/pii/B9780128192405000031?via%3Dihub.

Kasar PM, Tribhuwan CS, Khode JG. 2020. Innovative packaging of medicines. *Asian J. Res. Pharm. Sci.* 10 (1), 56–60. https://doi.org/10.5958/2231-5659.2020.00011.9.

Kawalkar A. 2014. A comprehensive review on osteoporosis. *J. Trauma Orthop.* 9 (4), 3–12.

Keenan TR. 2012. Gelatin. *Polym. Sci. A Compr. Ref.* 10, 237–247. https://doi.org/10.1016/B978-0-444-53349-4.00265-X.

Kommareddy S, Shenoy DB, Amiji MM. 2007. Gelatin nanoparticles and their biofunctionalization. *Nanotechnol. Life Sci.* 1, 330–352. https://doi.org/10.1002/9783527610419.ntls0011.

Kumar ABV, Gowda LR, Tharanathan RN. 2004. Non-specific depolymerization of chitosan by pronase and characterization of the resultant products. *Eur. J. Biochem.* 271 (4), 713–723. https://doi.org/10.1111/j.1432-1033.2003.03975.x.

Kumari S, Rath PK. 2014. Extraction and characterization of chitin and chitosan from (Labeo Rohit) fish scales. *Procedia Mater. Sci.* 6, 482–489. https://doi.org/10.1016/j.mspro.2014.07.062.

Le Bourlais C, Acar L, Zia H, Sado PA, Needham T, Leverge R. 1998. Ophthalmic drug delivery systems—Recent advances. *Prog. Retin. Eye Res.* 17 (1), 33–58. https://doi.org/10.1016/S1350-9462(97)00002-5.

Li MC, Wu Q, Song K, Cheng HN, Suzuki S, Lei T. 2016. Chitin nanofibers as reinforcing and antimicrobial agents in carboxymethyl cellulose films: Influence of partial deacetylation. *ACS Sustain. Chem. Eng.* 4 (8), 4385–4395. https://doi.org/10.1021/acssuschemeng.6b00981.

Li Q, Dunn ET, Grandmaison EW, Goosen MFA. 1992. Applications and properties of chitosan. *J. Bioact. Compat. Polym.* 7 (4), 370–397. https://doi.org/10.1177/088391159200700406.

Liu LS, Fishman ML, Kost J, Hicks KB. 2003. Pectin-based systems for colon-specific drug delivery via oral route. *Biomaterials* 24 (19), 3333–3343. https://doi.org/10.1016/S0142-9612(03)00213-8.

Liu Y, Ahmed S, Sameen DE, Wang Y, Lu R, Dai J, Li S, Qin W. 2021. A review of cellulose and its derivatives in biopolymer-based for food packaging application. *Trends Food Sci. Technol.* 112, 532–546. https://doi.org/10.1016/j.tifs.2021.04.016.

Liu Y, Gan L, Carlsson DJ, Fagerholm P, Lagali N, Watsky MA, Munger R, Hodge WG, Priest D, Griffith M. 2006. A simple, cross-linked collagen tissue substitute for corneal implantation. *Investig. Ophthalmol. Vis. Sci.* 47 (5), 1869–1875. https://doi.org/10.1167/iovs.05-1339.

Martău GA, Mihai M, Vodnar DC. 2019. The use of chitosan, alginate, and pectin in the biomedical and food sector-biocompatibility, bioadhesiveness, and biodegradability. *Polymers* 11 (11), 1837. https://doi.org/10.3390/polym11111837.

Michel AS, Mestdagh MM, Axelos MAV. 1997. Physico-chemical properties of carrageenan gels in presence of various cations. *Int. J. Biol. Macromol.* 21, 195–200. https://doi.org/10.1016/S0141-8130(97)00061-5.

Mohammed MH, Williams PA, Tverezovskaya O. 2013. Extraction of chitin from prawn shells and conversion to low molecular mass chitosan. *Food Hydrocoll.* 31 (2), 166–171. https://doi.org/10.1016/j.foodhyd.2012.10.021.

Mohan S, Oluwafemi OS, Kalarikkal N, Thomas S, Songca SP. 2016. Biopolymers—Application in nanoscience and nanotechnology. In: *Recent Advances in Biopolymers*, Parveen FK (Ed.), IntechOpen, Rijeka, Croatia. https://doi.org/10.5772/62225. https://www.intechopen.com/chapters/49884.

Moslemi M. 2021. Reviewing the recent advances in application of pectin for technical and health promotion purposes: From laboratory to market. *Carbohydr. Polym.* 254 (24), 117324. https://doi.org/10.1016/j.carbpol.2020.117324.

Munarin F, Tanzi MC, Petrini P. 2012. Advances in biomedical applications of pectin gels. *Int. J. Biol. Macromol.* 51 (4), 681–689. https://doi.org/10.1016/j.ijbiomac.2012.07.002.

Muñoz I, Rodríguez C, Gillet D, Moerschbacher B. 2018. Life cycle assessment of chitosan production in India and Europe. *Int. J. Life Cycle Assess.* 23 (5), 1151–1160. https://doi.org/10.1007/s11367-017-1290-2.

Musgrave CSA, Fang F. 2019. Contact lens materials: A materials science perspective. *Materials* 12 (2), 1–35. https://doi.org/10.3390/ma12020261.

Nagarajan S, Radhakrishnan S, Kalkura SN, Balme S, Miele P, Bechelany M. 2019. Overview of protein-based biopolymers for biomedical application. *Macromol. Chem. Phys.* 2020 (14), 1900126.

Necas J, Bartosikova L. 2013. Carrageenan: A review. *Vet. Med.* 58 (4), 187–205. https://doi.org/10.17221/6758-VETMED.

Nedovic V, Kalusevic A, Manojlovic V, Levic S, Bugarski B. 2011. An overview of encapsulation technologies for food applications. *Procedia Food Sci.* 1, 1806–1815. https://doi.org/10.1016/j.profoo.2011.09.265.

Pal K, Banthia AK, Majumdar DK. 2006. Preparation of transparent starch based hydrogel membrane with potential application as wound dressing. *Trends Biomater. Artif. Organs* 20 (1), 59–67.

Pereira CS, Cunha AM, Reis RL, Vázquez B, San Román J. 1998. New starch-based thermoplastic hydrogels for use as bone cements or drug-delivery carriers. *J. Mater. Sci.: Mater. Med.* 9, 825–833. https://doi.org/10.1023/A:1008944127971.

Premraj R, Doble M. 2005. Biodegradation of polymers. *Indian J. Biotechnol.* 4(2), 186–193. https://doi.org/10.17516/1997-1389-2015-8-2-113-130.

Rashmi SH, Mahendra, Biradar B, Maladkar K, Kittur AA. 2016. Extraction of chitin from prawn shell and preparation of chitosan. *Res. J. Chem. Environ. Sci.* 4, 70–73.

Rathore KS, Nema RK. 2009. Review on ocular inserts. *Int. J. PharmTech Res.* 1 (2), 164–169.

Ricard-Blum S. 2011. The collagen family. Cold Spring Harb. *Perspect. Biol.* 3 (1), a004978. https://doi.org/10.1101/cshperspect.a004978.

Rinaudo M. 2006. Chitin and chitosan: Properties and applications. *Prog. Polym. Sci.* 31 (7), 603–632. https://doi.org/10.1016/j.progpolymsci.2006.06.001.

Riswati SS, Setiati R, Kasmungin S, Fathaddin MT. 2021. Current development of cellulose-based biopolymer as an agent for enhancing oil recovery. *IOP Conf. Ser. Earth Environ. Sci.* 802 (1), 102023. https://doi.org/10.1088/1755-1315/802/1/012023.

Rodrigues A, Emeje M. 2012. Recent applications of starch derivatives in nanodrug delivery. *Carbohydr. Polym.* 87 (2), 987–994. https://doi.org/10.1016/j.carbpol.2011.09.044.

Ruszczak Z, Friess W. 2003. Collagen as a carrier for on-site delivery of antibacterial drugs. *Adv. Drug Deliv. Rev.* 55 (12), 1679–1698. https://doi.org/10.1016/j.addr.2003.08.007.

Salah R, Michaud P, Mati F, Harrat Z, Lounici H, Abdi N, Drouiche N, Mameri N. 2013. Anticancer activity of chemically prepared shrimp low molecular weight chitin evaluation with the human monocyte leukaemia cell line, THP-1. *Int. J. Biol. Macromol.* 52 (1), 333–339. https://doi.org/10.1016/j.ijbiomac.2012.10.009.

Santos VP, Marques NSS, Maia PCSV, de Lima MAB, de Oliveira Franco L, de Campos-Takaki GM. 2020. Seafood waste as attractive source of chitin and chitosan production and their applications. *Int. J. Mol. Sci.* 21 (12), 4290. https://doi.org/10.3390/ijms21124290.

Schmitz C, Auza LG, Koberidze D, Rasche S, Fischer R, Bortesi L. 2019. Conversion of chitin to defined chitosan oligomers: Current status and future prospects. *Mar. Drugs* 17 (8), 452. https://doi.org/10.3390/md17080452.

Seetharaman K, Bertoft E. 2012. Perspectives on the history of research on starch: Part I: On the linkages in starch. *Starch/Staerke* 64 (9), 677–682. https://doi.org/10.1002/star.201200088.

Shahrin EWES, Narudin NAH, Padmosoedarso KM, Kusrini E, Mahadi AH, Shahri NNM, Usman A. 2021. Pectin derived from pomelo pith as a superior adsorbent to remove toxic Acid Blue 25 from aqueous solution. *Carbohydr. Polym. Technol. Appl.* 2, 100116. https://doi.org/10.1016/j.carpta.2021.100116.

Shankar S, Rhim J-W. 2018. Bionanocomposite films for food packaging applications. In: *Reference Module in Food Science*, Smithers G, Trinetta V, Knoerzer K (Eds.), Elsevier, Amsterdam, Netherlands, pp. 1–10. https://doi.org/10.1016/b978-0-08-100596-5.21875-1. https://www.elsevier. com/books/bionanocomposites-for-food-packaging-applications/ahmed/978- 0-323-88528-7.

Shrivastava A. (2018). Introduction to plastics engineering. In: *Introduction to Plastics Engineering*, Shrivastava A. (Ed.), Elsevier, Amsterdam, Netherlands, pp. 1–16. https://doi.org/10.1016/b978-0-323-39500-7.00001-0. https://www.elsevier.com/books/introduction-to-plastics-engineering/shrivastava/978-0-323-39500-7.

Sionkowska A, Skrzyński S, Śmiechowski K, Kołodziejczak A. 2017. The review of versatile application of collagen. *Polym. Adv. Technol.* 28 (1), 4–9. https://doi.org/10.1002/pat.3842.

Sivakanthan S, Rajendran S, Gamage A, Madhujith T, Mani S. 2020. Antioxidant and antimicrobial applications of biopolymers: A review. *Food Res. Int.* 136, 109327. https://doi.org/10.1016/j.foodres.2020.109327.

Sriamornsak P. 2011. Application of pectin in oral drug delivery. *Expert Opin. Drug Deliv.* 8 (8), 1009–1023. https://doi.org/10.1517/17425247.2011.584867.

Timilsena YP, Haque MA, Adhikari B. 2020. Encapsulation in the food industry: A brief historical overview to recent developments. *Food Nutr. Sci.* 11 (06), 481–508. https://doi.org/10.4236/fns.2020.116035.

Udayakumar GP, Muthusamy S, Selvaganesh B, Sivarajasekar N, Rambabu K, Banat F, Sivamani S, Sivakumar N, Hosseini-Bandegharaei A, Show PL. 2021a. Biopolymers and composites: Properties, characterization and their applications in food, medical and pharmaceutical industries. *J. Environ. Chem. Eng.* 9 (4), 105322. https://doi.org/10.1016/j.jece.2021.105322.

Udayakumar GP, Muthusamy S, Selvaganesh B, Sivarajasekar N, Rambabu K, Sivamani S, Sivakumar N, Maran JP, Hosseini-Bandegharaei A. 2021b. Ecofriendly biopolymers and composites: preparation and their applications in water-treatment. *Biotechnol. Adv.* 52, 107815. https://doi.org/10.1016/j.biotechadv.2021.107815.

Varun TK, Senani S, Jayapal N, Chikkerur J, Roy S, Tekulapally VB, Gautam M, Kumar N. 2017. Extraction of chitosan and its oligomers from shrimp shell waste, their characterization and antimicrobial effect. *Vet. World* 10 (2), 170–175. https://doi.org/10.14202/vetworld.2017.170–175.

Venkateshaiah A, Padil VVT, Nagalakshmaiah M, Waclawek S, Černík M, Varma RS. 2020. Microscopic techniques for the analysis of micro and nanostructures of biopolymers and their derivatives. *Polymers* 12 (3), 512. https://doi.org/10.3390/polym12030512.

Verma S, Manjubala I, Narendrakumar U. 2016. Protein and carbohydrate biopolymers for biomedical applications. *Int. J. PharmTech Res* 9 (8), 408–421.

Vongchan P, Sajomsang W, Kasinrerk W, Subyen D, Kongtawelert P. 2003. Anticoagulant activities of the chitosan polysulfate synthesized from marine crab shell by semi-heterogeneous conditions. *ScienceAsia* 29 (2), 115–120. https://doi.org/10.2306/scienceasia1513-1874.2003.29.115.

Wankhade V. 2020. Animal-derived biopolymers in food and biomedical technology. In: *Biopolymer-Based Formulations*, Pal K, Banerjee I, Sarkar P, Kim D, Win-Ping Deng W-P, Dubey NK, Majumder K. (Eds.), Elsevier Inc., Amsterdam, Netherlands, pp. 135–152. https://doi.org/10.1016/B978-0-12-816897-4.00006-0. https://www.sciencedirect.com/science/article/pii/B9780128168974000060.

Wong TW, Colombo G, Sonvico F. 2011. Pectin matrix as oral drug delivery vehicle for colon cancer treatment. *AAPS PharmSciTech* 12 (1), 201–214. https://doi.org/10.1208/s12249-010-9564-z.

Yadav P, Yadav H, Shah VG, Shah G, Dhaka G. 2015. Biomedical biopolymers, their origin and evolution in biomedical sciences: A systematic review. *J. Clin. Diagnostic Res.* 9 (9), ZE21–ZE25. https://doi.org/10.7860/JCDR/2015/13907.6565.

Yen MT, Yang JH, Mau JL. 2008. Antioxidant properties of chitosan from crab shells. *Carbohydr. Polym.* 74 (4), 840–844. https://doi.org/10.1016/j.carbpol.2008.05.003.

Zadbuke N, Shahi S, Gulecha B, Padalkar A, Thube M. 2013. Recent trends and future of pharmaceutical packaging technology. *J. Pharm. Bioallied Sci.* 5 (2), 98–110. https://doi.org/10.4103/0975-7406.111820.

10 Conclusion and Future Prospects

Samsul Ariffin Abdul Karim
and Poppy Puspitasari

CONTENTS

10.1 CONCLUSION

Advanced materials will play a key role in the future of energy. These materials, which can include new types of metals, alloys, and composites, have the potential to improve the efficiency and performance of energy technologies, such as solar panels and batteries. For example, materials that are more conductive or have higher thermal conductivity could be used to improve the efficiency of photovoltaic cells, while new types of lightweight, high-strength materials could be used to make stronger and more durable wind turbine blades. Additionally, advanced materials that are able to withstand harsh environments, such as high temperatures or corrosive conditions, could be used in a variety of energy applications, including in fossil fuel power plants and nuclear reactors.

10.2 FUTURE PROSPECTS

There are many potential advanced materials that could be used for energy storage and capacitors in the future. For example, graphene, a two-dimensional form of carbon with exceptional conductivity and strength, has been proposed as a potential material for high-capacity, high-performance energy storage devices. Other advanced materials that are being explored for energy storage include nanostructured metals, alloys, and composites, as well as novel polymers and ceramics. These materials may have unique electrical, thermal, or mechanical properties that make them well-suited for energy storage applications. Additionally, new manufacturing techniques, such as 3D printing, could be used to create complex and customized energy storage devices using advanced materials

Agriculture sustainability refers to the practice of producing food and fiber in a way that maintains or improves the natural environment and the resources that are used in agricultural production. This can include practices such as using

DOI: 10.1201/9781003367819-10

sustainable farming methods, conserving water and soil resources, protecting biodiversity, and reducing the use of pesticides and other harmful chemicals. The goal of agriculture sustainability is to ensure that the food and fiber needs of the present population can be met without compromising the ability of future generations to meet their own needs. By adopting sustainable agriculture practices, farmers and other stakeholders in the agricultural industry can help to preserve the natural resources that are essential for food production and protect the environment for future generations

Photovoltaics will play a significant role in the future of energy sustainability. Photovoltaic technology, which converts sunlight into electricity, has become increasingly affordable and efficient in recent years, making it a viable option for generating clean, renewable energy. As the demand for sustainable energy sources continues to grow, it is likely that photovoltaic technology will become even more widespread and advanced. In the future, we may see the development of new types of photovoltaic materials and technologies that are able to convert a wider spectrum of sunlight into electricity, as well as the integration of photovoltaic systems into the design of buildings and infrastructure. Overall, the future of photovoltaics in energy sustainability looks bright, especially to cater to the objectives of SDGs [1] and carbon net zero by the year 2050.

10.3　ACKNOWLEDGMENTS

Special thank-you to the Faculty of Computing and Informatics, Universiti Malaysia Sabah, for the financial and computing facilities support that has made the completion of the book possible.

REFERENCE

[1] https://sdgs.un.org/goals (Retrieved on 22 December 2022).

Index

Printed in the United States
by Baker & Taylor Publisher Services